Long-Term Strength of Materials

The physics of fracture processes, which includes Fracture mechanics, is crucial for understanding the longevity and reliability of any structure, from fracture initiation to propagation and final catastrophic failure. This textbook introduces the thermodynamics of irreversible processes along with entropy to address the time dependency of fracture.

Working from observations of structural failure, the book identifies the principal failure types such as brittle fracture, with considerations of solo crack initiation and crack propagation associated with collective distributed damage. The other type is ductile fracture, when a crack blunts immediately on the application of stress resulting in large deformation. The book then addresses the life of a structure in a specific environment and load condition, using irreversible thermodynamics and the entropy criterion to address cooperative fracture and novel statistical Fracture mechanics to address solo fracture.

- Applies well-established concepts from mechanics, absent in contemporary Fracture mechanics
- Uses novel concepts of mechanics, irreversible thermodynamics, and statistical Fracture mechanics

The book is ideal for graduate students and design engineers in civil and materials engineering, as well as mechanical and chemical engineering. Students using the book need no more than basic college-level mechanics, mathematics, and statistics knowledge.

Alexander Chudnovsky is UIC Distinguished Professor Emeritus and Director of the Fracture Mechanics and Materials Durability Laboratory in the Civil and Materials Engineering Department of the University of Illinois at Chicago, USA. Over a career of more than 60 years, he has received numerous teaching awards and the 2004 Outstanding Achievement Award of the Society of Plastics Engineers.

Kalyan Sehanobish has been a global plastic product consultant and owner of Materials & Adhesives Research Services since 2019, and for over 25 years was Research Fellow at the Dow Chemical Company prior to 2019. He received the R&D inventor of the year award for several projects as a technical project team participant and the Automotive PACE award for IMPAXX foam in 2008.

Long-Term Strength of Materials

Reliability Assessment and Lifetime Prediction of Engineering Structures

Alexander Chudnovsky and
Kalyan Sehanobish

CRC Press
Taylor & Francis Group
Boca Raton London New York

CRC Press is an imprint of the
Taylor & Francis Group, an **informa** business

Designed cover image: © Alexander Chudnovsky

First edition published 2024
by CRC Press
6000 Broken Sound Parkway NW, Suite 300, Boca Raton, FL 33487–2742

and by CRC Press
4 Park Square, Milton Park, Abingdon, Oxon, OX14 4RN

CRC Press is an imprint of Taylor & Francis Group, LLC

© 2024 Alexander Chudnovsky and Kalyan Sehanobish

Library of Congress Cataloging-in-Publication Data
Names: Chudnovsky, Alexander, author. | Sehanobish, Kalyan, author.
Title: Long-term strength of materials : reliability assessment and lifetime
    prediction of engineering structures / Alexander Chudnovsky and Kalyan
    Sehanobish.
Description: First edition. | Boca Raton, FL : CRC Press, 2024. |
    Includes bibliographical references and index.
Identifiers: LCCN 2023008757 | ISBN 9781032418131 (paperback) |
    ISBN 9781032418148 (hardback) | ISBN 9781003359845 (ebook)
Subjects: LCSH: Strength of materials. | Service life (Engineering)
Classification: LCC TA405 .C535 2024 |
    DDC 620.1/12—dc23/eng/20230323
LC record available at https://lccn.loc.gov/2023008757

ISBN: 978-1-032-41814-8 (hbk)
ISBN: 978-1-032-41813-1 (pbk)
ISBN: 978-1-003-35984-5 (ebk)

DOI: 10.1201/9781003359845

Typeset in Sabon
by Apex CoVantage, LLC

Access the Support Material: www.routledge.com/9781032418131

An old man, going a lone highway,
Came, at the evening, cold and gray,
To a chasm, vast, and deep, and wide,
Through which was flowing a sullen tide.
The old man crossed in the twilight dim;
The sullen stream had no fears for him;
But he turned, when safe on the other side,
And built a bridge to span the tide.

"Old man," said a fellow pilgrim, near,
"You are wasting strength with building here;
Your journey will end with the ending day;
You never again must pass this way;
You have crossed the chasm, deep and wide-
Why build you a bridge at the eventide?"

The builder lifted his old gray head:
"Good friend, in the path I have come," he said,
"There followeth after me today,
A youth, whose feet must pass this way.
This chasm, that has been naught to me,
To that fair-haired youth may a pitfall be.
He, too, must cross in the twilight dim;
Good friend, I am building the bridge for him."
Author: Will Allen Dromgoole

We write this book for our grandsons and their grandsons who would eventually need some of the tools shared in this book. We do not know what the future holds but contemporary society is overwhelmed by technological leads, but fundamental science is being ignored. Just as any other field, the field of mechanics has reached a saturation that needs new fundamentals to advance technology beyond the bounds of earth, our current home.

Contents

Preface

Fracture mechanics is the natural extension of the traditional strength of materials course for most of the engineering professions. The strength of materials got the reputation of a difficult engineering course probably due to mathematical formalism it operates with, more specifically the tensorial algebra which is something new for any graduate from a regular high school. However, the concept of the tensors and tensorial algebra are very natural extensions of basic algebra and could be easily understood if explained in a proper order. Moreover, such concepts and formalism are necessary for the purpose of adequately addressing the issue of longevity and reliability of any structure (engineering, natural, and cosmic). It is the only precise language to describe the multidimensional world around us.

But one of the shortcomings of traditional strength of materials course is that it does not explicitly address time. In this book, we are going to depart from the traditional approach of strength of materials and Fracture mechanics and introduce the thermodynamics of irreversible processes to address the time dependency of fracture. To contemplate, we need to first summarize the observed facts associated with failure of structures in service. The next step is to identify the principal failure types such as brittle (solo crack initiation and growth versus collective distributed damage, "Crack layer", ahead of the crack also can be attributed as cooperative fracture) and ductile (when crack blunts immediately on the application of stress resulting in a large deformation, e.g., tearing in thin plastic film). But then to complete it, we must address lifetime in a specific environment and load condition. While irreversible thermodynamics, specifically *entropy criterion*, is necessary to address cooperative fracture, a novel *statistical Fracture mechanics* (SFM) concept will be introduced to address solo fracture.

Crack propagator of the SFM can be visualized as a probability cloud hanging above a domain of concentration of the crack-tip stress field ahead of the crack [Chudnovsky (1973), Chudnovsky and Kunin (1987, 1992), Chudnovsky, A. and Gorelik, M. (1994, 1997)]. In the case of cooperative fracture, the probability cloud [Chudnovsky, A. and Kunin, B. (1987, 1992)] is materialized as a damage zone which is also called the Process Zone (PZ). It consists of microscopic and submicroscopic defects formed due to the high stress induced by the crack. A system of crack and the PZ is named Crack layer (CL). CL evolution is discussed in earlier publications [Chudnovsky, A. (2014); Chudnovsky and Kunin, (1987)]. We are writing this book for our grandsons and grand grandsons who will face a challenge of designing the engineering structures intended to operate at the extreme ambient conditions, which presently we do not see here on the earth under the protection of the earth's atmosphere.

At this point, I wish to extend my thanks to all my students who helped me experimentally examine these theoretical ideas which would have remained a figment of my imagination in former Soviet Union. Students who are worth mentioning are Dr. John Botsis for performing the first experiments on polystyrene sheets to demonstrate the utility of Crack layer theory, Dr. Haris Jasarevic for theoretically and experimentally addressing the most avoided topic of crack initiation, Dr. Boris Kunin for helping with the development of statistical fracture theory, and Dr. Haying Zhang for providing multiple technical contributions throughout the development of these ideas and compilation of this book. Finally, thanks are due to Dr. Kalyan Sehanobish for compiling my life's work. I must sincerely thank two important ladies in life: my wife Elizabeth Chudnovsky and my sister-in-law Yelena Kurdina who were instrumental in giving me all the support I needed for my sustenance through my good and adverse health conditions.

Alexander Chudnovsky
November 2021

I would like to thank my wife Suprita C. Sehanobish for volunteering and helping me to do this work over breakfast at Prof. Chudnovsky's place. Whatever I know about critical thinking and materials science is primarily due to professor's guidance. This was a tough task since professors' lifetime of work is not easy to comprehend and required some serious learning. I am not sure whether I have done my job right to his satisfaction but certainly have tried my best.

Kalyan Sehanobish
July 2022

Introduction

From the time mankind began building engineering structures, they always had to struggle with the topic of "lifetime" of these structures [lifetime addressed does not include any unnatural situations]. If we go as far as 1750 BCE (Before Current Era) during the time of Hammurabi, historians discovered the rules of construction that require the builder to be put to death if the house he built collapsed killing the owner (Figure 0.1). Builder's son was to be put to death in father's absence. Such harsh laws required the builder to be extra careful. Since then, we learnt a lot about designing structures, lifetime prediction modelling, engineering analysis, and Fracture mechanics leading to more complex structures. However, we still cannot predict lifetime accurately enough due to huge gaps in our knowledge. Severe penalties associated with structural failure in Hammurabi's time (Figure 0.1) are replaced by the modern form of penalties, but even today's builders have no recourse other than depending on safety factors. To address these knowledge gaps, we must look at the problem of failure in a more inclusive manner beyond human engineering achievements alone. We will soon learn that simple approaches as reducing stress on the structure do not always lead to safe structures.

Cosmic Fracture Phenomena

Fracture phenomena is observable in the formation of planetary systems within the universe (Figure 0.2). The universe underwent a single fragmentation event, separating into protogalactic volumes at a relatively early stage after the Big Bang, which was discovered on the basis of the astrophysical findings. A model of fragmentation of the matter within the universe in the process of expansion was first proposed by Grady et al. Joe Grady was Chief of the Ceramic and Polymer Composites Branch at NASA Glenn Research Center in Cleveland, the United States. After receiving a Ph.D. degree from the Department of Aeronautics and Astronautics at Purdue University, he conducted research on failure mechanics of high-temperature engine materials. Grady once derived a simple model of dynamic fragmentation within the universe by balancing the

DOI: 10.1201/9781003359845-1

Laws of Hammurabi (1792 - 1750 BCE)

The codes #228 through #233 of the Hammurabi's Code of Laws (282 Codes) represent the rules of construction in Babylonia of that time.

229 If a builder has built a house for a man and has not made his work sound, and the house he built has fallen, causing the death of its owner, that builder shall be put to death.

Figure 0.1 Hammurabi Codes declaring the punishment of the builder and his future generations in the case of the builder's absence.

available kinetic energy against the energy associated with the new surface created in the process. This model neglected the effect of stored elastic (strain) energy, and a major correction was made by Chudnovsky (referred as AC) et al. by adding gravitational energy as potential energy. This resulted in a prediction of fragment size as a function of expansion rate at the later stage of universe expansion. This was subsequently supported by experimental data from Army Research laboratories [Brown et al. (1983), Glenn et al. (1986)]. Private communications with researchers [Brown et al. (1983)] clearly established the power of modelling and prediction of the fragment size by Chudnovsky et al. Glenn–Chudnovsky model predicts that the large fragments are formed at a low stress, and the smallest fragments are formed at the onset of the macroscopic fracture under high stress. If the average fragment size is determined primarily by the kinematics of crack growth, the theory of linear elastic Fracture mechanics (LEFM) can be employed to clarify matters. Crack nuclei are supposed to be homogeneously distributed within the body, and, after initiation, the cracks are assumed to propagate without branching, rotation, or kinking, so that the crack length is the only parameter characterizing propagation (and, eventually, fragment size). Moreover, the crack walls are taken to remain free of surface tractions. We specifically exclude any cracks that form by wavefront coalescence. Energy conservation then requires that the total kinetic and potential energy release associated with crack propagation is spent partially on the formation of new surface and is partially dissipated.

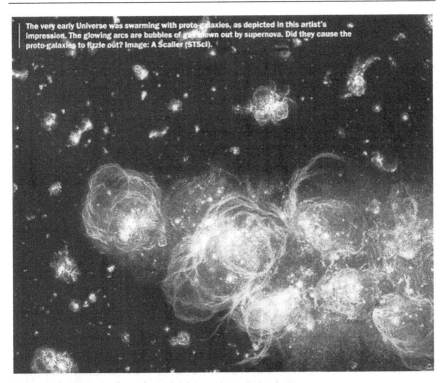

The very early Universe was swarming with proto-galaxies, as depicted in this artist's impression. The glowing arcs are bubbles of gas blown out by supernova. Did they cause the proto-galaxies to fizzle out? Image: A Scalier (STScI).

The Universe within 50000 Light Years: The Milky Way Galaxy

Figure 0.2 Fragmentation of the universe.

Considering incremental surface energy as fragmentation occurs and using the fundamental assumption that the fragmentation process occurs with minimal dissipation, one can predict the fragment size. This is based on the presumption that crack speeds are small compared with wave speeds and hence that the system remains in thermodynamic equilibrium with its surrounding as the cracks extend. From LEFM theory, the energy release per unit crack area for (opening mode) can be derived including the Rayleigh wave speed in the equation. Homogeneous dilatation of the medium leads to an increase of stress during the process of crack growth. Thus, one can reach a solution for fragment radius employing a strain-energy-dominant solution. Classical Fracture mechanics is incapable of addressing such cosmic phenomenon. In subsequent chapters, we will bring forward these new predictive capabilities offered by the authors of this book.

Fracture Phenomena in Nature

Fracture phenomena are observable in earthquakes/volcanic eruptions, avalanches, gas/oil extractions, and landslides that frequently occur in our blue planet, earth. We notice that lithosphere is essentially made of *tectonic plates* (pieces of the Earth's crust) and uppermost mantle. There are 52 major (14) and minor (38) plates of about 100 km (60 miles) thickness, and they consist of two principal types of material: ocean crust (Si and Mg—sima) and continental crust (Si and Al—sial). Epicenters of 3,58,214 earthquake events that occurred between 1963 and 1998 are represented by black dotted regions and are distributed primarily along the tectonic plate's weak boundaries pointing out that this is also a fracture phenomenon (Figure 0.3).

Figure 0.4 is a schematic illustrating the centers of deformation and fracture along the rift valley of the lithosphere. *Lithosphere* is the solid part of the earth and consists of three main layers—crust, mantle, and core. A rift valley is a lowland region that forms where earth's tectonic plates move apart. These valleys are found both on land and at the bottom of the ocean where they are created by the process of seafloor spreading. Picture also shows the *Asthenosphere*, the upper layer of the earth's mantle, below the lithosphere.

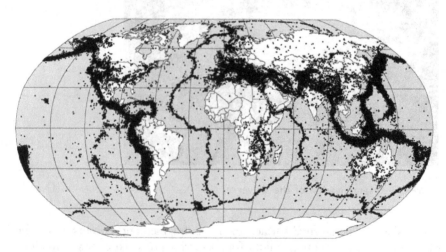

Figure 0.3 Distribution of epicenters of earthquakes.

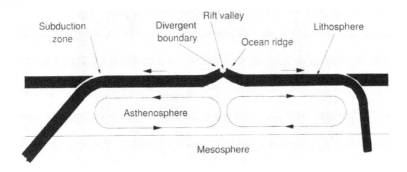

Figure 0.4 Lithosphere deformation and fracture.

Ocean ridge is a seafloor mountain system formed by plate tectonics and is associated with seafloor spreading. *Subduction zone* shown in the picture is a region of the earth's crust where tectonic plates meet. A *divergent boundary* occurs when two tectonic plates move away from each other. Along these boundaries, earthquakes are common, and magma (molten rock) rises from earth's mantle to the surface, solidifying to create new oceanic crust.

Avalanches in snow-covered landscapes also initiate at a fault line within the snow accumulation. Figure 0.5 shows one such avalanche in Lionhead Ridge several years ago. The man in the picture is pointing to the fault line. To understand avalanches and take some preventive measures, one needs to recall that snow is a complex three-phase system that exists in the dynamic equilibrium of constant exchanges of the solid, liquid, and gaseous states. The phase transitions from the solid state (ice into water), and possible evaporation may depend on the random fluctuations of the temperature, partial pressure, and the roughness of the substrate such as rocks, grass, and/or dirt as well as the air motion. At certain sites, this process forms a very hard layer or lens-shaped section (referred to as ice lens) in a snowpack of solid or near solid ice known as ice lens. Ice lenses are bodies of ice formed when moisture, diffused within soil or rock, accumulates in a localized zone. The ice initially accumulates within small, collocated pores or pre-existing crack, and, if the conditions remain favorable, continues to collect in the ice layer or ice lens, wedging the soil or rock apart. Ice lenses grow parallel to the surface and several centimeters to several decimeters (inches to feet) deep in the soil or rock. Studies have demonstrated that rock fracture happens by ice segregation (i.e., the fracture of intact rock by ice lenses that grow by drawing water from their surroundings during periods of sustained subfreezing temperatures). Ice lens on the substrate may trigger the sliding (mode II type fracture explained in Chapter IV) movement of the heavy layers of the wet snow above the lens. Monitoring the state of the snow would allow making assessment of the chances for triggering avalanche, usually done by experienced people.

Fracture in Engineering Structures

Extraction of gas and oil takes advantage of layered structure and preexisting fractures of rocks. Hydraulic fracturing (HF) is used to enhance the

LIONHEAD RIDGE AVALANCHE
26 January

32 degrees

Avalanche slid on Surface Hoar
layer 1 foot deep

Figure 0.5 Avalanches and a man pointing at the origin of the avalanche.

production. There is a natural layering in sedimentary rocks resulting from the history of formation (Figures 0.6–0.12). HF is an enormous effort, and it needs quality fracture data before finalizing a strategy. Figure 0.6 is a

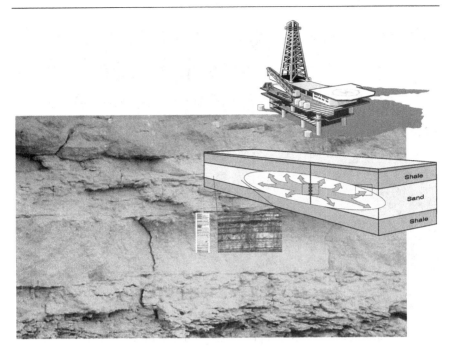

Figure 0.6 Challenges in using natural gas in oil exploration and production.

schematic of an offshore platform commonly used to drill deep well under the bottom of an ocean for oil and natural gas production. Missing in the picture are the anchors that keep it in the steady position preventing a drift due to inevitable wind and currents. The schematic also shows typical layered structures in the rocks at the bottom of the ocean. Such offshore platforms are used to inject some special liquids (primarily diluted hydrochloric acid) to produce hydraulic fracture as well as for injection of water to recover the elasticity of the rocks that undergo softening with the gas production which may cause subsiding of the entire region. This is important for countries like the Netherlands which is the number 5 exporter of natural gas today, and the same country is in most part below the sea level. Thus, the country is also protected from collapse by dams erected along the shore.

Figure 0.7 illustrates layered structure of the rocks and preexisting fracture within the rocks. Figure 0.8 highlights a portion of the hydraulic fracture in nature. To extract fracture data, field failure testing is normally employed. This is a very expensive large-scale endeavor. Figure 0.9 illustrates how field testing is done to collect information about hydraulic fractures. For field testing, holes are drilled and the liquid is injected with a special pump into the rocks to create a hydraulic fracture as it is done in actual oil and gas drilling operations. Various sensors are attached to the surrounding area to measure tilt accurately as the fracture is induced. Once the fracture is induced,

Figure 0.7 Layered structures and preexisting fracture of rocks (taken by AC).

Figure 0.8 Hydraulic fracture in nature (taken by AC).

Figure 0.9 Field testing of hydraulic fracture.

one goes below the ground to unearth the hydraulic fracture. Figure 0.10 shows the exposed fracture site, and Figure 0.11 clearly reveals the details of the fracture path. After identifying the fracture, Fracture mechanics parameters like crack opening displacement (COD) are obtained to correlate with the imposed pressure on the rocks. As shown in Figure 0.12, the width of solidified fracturing liquid (white layer) is measured as a representation of COD. This type of field testing is very expensive to do and provides a limited amount of data for developing fracture predictive models, which needs multiple repetition.

Figures 0.13 and 0.14 taken in Wyoming show the scale of operation of coal excavation and transportation. The same rock layer that contains the coal in these parts has moved several miles away down to a depth of about 4,500 ft below the surface toward the center of the earth. Royal Dutch Shell (currently Shell plc.— British company) made a deep bore hole to pump natural gas. When gas production started to decline, appropriate hydraulic fracture methods were considered. But to do that one needs to know the basic mechanical properties of the rock. However, it is hard to get sufficiently large specimens from the depth of a few thousand feet for testing. Fortunately, some geologists suggested that composition and structure of the rock on such depth could be the same within the same geological layer as it has risen due to the tectonic plate movement. This would allow one to collect a piece of rock at the site of open excavation as a representative to study composition,

Figure 0.10 Unearthing of hydraulic fracture.

Figure 0.11 Fracture revealed.

Figure 0.12 Measuring COD of hydraulic fracture.

Figure 0.13 Big trucks (50 T load-bearing capacity) used to transport the coal in a site in Wyoming. Prof. Chudnovsky is pointing to the road sign to give an idea of scale of operation.

Figure 0.14 Prof. Chudnovsky pointing to the driver sitting high up in the air with limited visibility immediately ahead of him.

grain size distribution, hardness, strength, and toughness of such a rock deep down. Doubts will remain about the loading and environmental history differences which need to be taken into consideration. Next, Professor Chudnovsky, as a consultant, developed a testing capability to mimic such field failure in University of Illinois Fracture laboratories with financial support from Shell and subsequently from Total (currently Total Energies).

Landslides also occur from a preexisting defect structure within uppermost mantle of earth due to a wide range of ground movements, such as rock falls, deep-seated slope failures, mudflows, and debris flows. Landslides occur in a variety of environments, characterized by either steep or gentle slope gradients: from mountain ranges to coastal cliffs or even underwater. Gravity is the primary driving force for a landslide to occur. Figures 0.15–0.18 show a closed road due to a collapsed bridge that connected the Riverside city, California, United States, with the external world. The local government has refused to start any reconstruction until legal investigation would be established to determine who was at fault and who would be responsible for the financial loses, since they suspected that the land sliding could be related to the premature failure of the polybutylene (PB) tubing failure in the region.

Figure 0.15 Landslide in Riverside, California, USA, which resulted in a broken bridge.

Figure 0.16 Section of landside adjacent to the broken bridge.

Figure 0.17 Cordoned area around the land slide adjacent to the highway.

Figure 0.18 Close-up view of the landslide.

Use of PB tubing in cold water distribution was quite common at that time of this incident.

Prof. Chudnovsky worked as technical expert witness for the court case. He found the actual failure site (he points to it in Figure 0.19) and established that the PB failure was not the cause of land sliding, since PB tubing was broken somewhere between the main and a water meter well. It appeared to be a pinhole failure with a small water leak. A water stream from the hole created a whirlpool with hard soil particles around the tubing. This caused erosion of the relatively soft tubing. However, the tubing failure looked like a clear break under tension with no sign of erosion commonly accompanied by a pinhole leak. This indicated that there was no leak prior to the ultimate PB tubing failure. Thus, the landsliding was caused by an excessive watering for landscaping purposes in the new subdivision constructed above that site and a multilayered soil slope with a high content of clay layers in the soil. Fracture surface inspection allowed one to make this conclusion. Although the professor was an expert for the local government, he maintained his claim with clear data. In most failure root cause finding, there is always some uncertainty, but quality analysis can be unrefutably disproven.

Now, let us look at some major failures of human-made engineering structures and root causes in the last 100 years. Bridges had been failing around the world as recently as a few months due to issues like negligence, poor

Figure 0.19 Prof. Chudnovsky is pointing to the actual failure origin.

inspection, and poor design. Worth mentioning is collapse of Florida International University–Sweetwater University City Bridge just five days after its installation in May 2018, killing at least six people, and the root cause is still unresolved. This is the sad reality of our ability to predict the lifetime of engineering structures. Figure 0.20 shows a fractured gusset plate recovered from the failure site. These plates are fastened to a permanent member either by bolts, rivets, or welding or a combination. They are made from thick sheets of steel to connect beams and columns together. It could be a possible root cause since lifetime was so short. Nothing was claimed in this case.

In 2007, I-35 Bridge collapsed in Minneapolis, Minnesota. The bridge collapsed during evening rush hour, and there was roadwork underway on the bridge. Main span of the bridge collapsed because the gusset plate fractured and missed regular inspection according to the published case study. In 1967, Point Pleasant Bridge collapsed due to a fractured eye bar [Bennett, A. and Mindlin, H. (1973)]. Figure 0.21 shows the debris and snippet of the original bridge. Figure 0.22 shows the fractured eye bar identified as the root cause of the collapse.

Failures are seen also in other types of structures such as catastrophes in the sea as shown in Figure 0.23 where a ship in mid sea splits into two halves for unreported causes. Similar failure in the sea occurred in a T2 tanker involved in World War II and an oil barge on transit in 1972 as shown in Figure 0.24 (a and b). In the case of the ship in Figure 0.23, details are missing. However, the failure is reminiscent of the failure of the famous ship Titanic. Although in the case of the Titanic, the failure was due to the collision with a giant iceberg, in the present case, no such accident was reported pointing to an internal failure. This type of extreme failures is not uncommon.

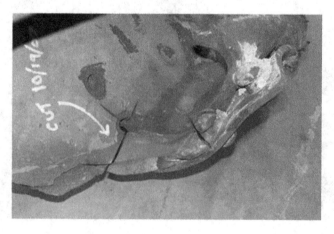

Figure 0.20 Fractured gusset plate.

Figure 0.21 *(a)* Point Pleasant bridge after its collapse in 1967 and *(b)* a similar collapse of the St. Mary's bridge.

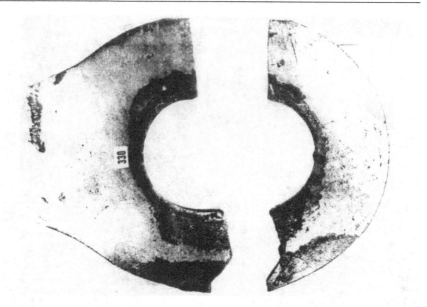

Figure 0.22 Fractured eye bar responsible for the failure of the Point Pleasant bridge.

Figure 0.23 A ship in mid sea splits into two halves due to a fracture event.

Figure 0.24 (a) Fractured T-2 tanker, the S.S. Schenectady in 1941 and (b) fractured oil barge, Martha R. Ingram, which failed on January 10, 1972 (New York Times).

Giant cargo ship, the *Crimson Polaris*, ran aground in a Japanese port and split in two. The ship suffered a crack that widened and eventually caused it to split in two pieces early morning on August 12, 2021. It's clear from this incident that one cannot blame an external agent for such failures. Failures like this could be avoided through meticulous inspection of these ships by experts in this profession. A significant number of splits in mid sea point to crack in the mid-ship region. However, all the identifications occur postmortem since crack identification and avoidance of splits of ships in mid sea will not be a newsworthy item.

On October 7, 1993, *New York Times National* published an article by Mathew W. I. Wald about a crack found in the core of a nuclear power plant (Figure 0.25). Figure 0.25 has key details of this article. Based on US Air Force Department of Defense, B1 bomber was found to be prematurely cracking due to metal fatigue. It was rammed through production, and the plane completed only 15% of its operational flight tests when it was declared operational in 1986; the cost of a B1 was $280 million. The cracks in the B1 bomber were found to be on the 25-degree longerons (load-bearing component of the framework), which were located to the left and right of

THE NEW YORK TIMES **NATIONAL** THURSDAY, OCTOBER 7, 1993

Crack Is Found Near the Core of a Nuclear Plant

By MATTHEW L. WALD

A steel cylinder that directs cooling water to the nuclear core of a North Carolina power plant has developed a crack around its 14-foot circumference, raising the possibility that it could break during an earthquake or other disaster and jam the mechanism that shuts the reactor, Federal officials say.

There are more than 30 reactors of similar design around the country, and the officials, at the Nuclear Regulatory Commission, said last week that they had warned these plants of the problem. It is the first time such a crack has been discovered in this kind of reactor, the officials said.

Separately, the Office of Technology Assessment, a Congressional agency, issued a report last week on aging nuclear power plants saying that "long-term prospects for the nation's 107 operating nuclear power plants are increasingly unclear."

Early Plant Retirements

Although plants are licensed for 40 years, no one really knows what a reactor's life expectancy is.

Six plants have been retired since 1989, all well before their operating licenses had expired, the study said, and the owners of several others are considering early closings. But the study said the recent history "may give a misleadingly dim view of the remaining lives of other nuclear power plants because of the great diversity among plants and changing electricity

30 reactors are warned after a disclosure in North Carolina.

market conditions."

In the case of the North Carolina reactor, Brunswick I, which is 17 years old. Federal officials have ordered the utility, the Carolina Power and Light Company, to keep the plant shut until it can repair the cylinder or prove that it cannot cause a major accident. The plant, in Southport, has been shut since April 1982 for refueling and improvements.

The crack runs more than half-way through the 3-inch-thick cylinder, called the core shroud, which is around the outside of the nuclear core but inside the reactor vessel, the giant steel pot in which water is boiled into steam for power.

The crack is apparently a result of years of radiation, repeated heating and cooling, unfavorable water chemistry in the reactor and stresses in the metal, said Stewart D. Ebneter, the regional administrator of the Nuclear Regulatory Commission in Atlanta. "It's a very complex mechanism," he said.

The utility found the crack after Gen-

eral Electric, which designed the plant, warned owners of similar plants that a crack had been found in a core shroud in Switzerland. But the crack in the Swiss plant was in a different part of the shroud, Mr. Ebneter said.

Inspection Finds More Cracks

Mr. Ebneter said that technicians examining the Brunswick shroud with ultrasonic gear found that "there are other cracks, but they are not as big or as deep, so far as we can tell." The inspection is complicated by the fact that the shroud is kept under water to hold down radiation levels.

In the worst case scenario, Mr. Ebneter said, the shroud could shift, bending parts inside the core, which might prevent operators from inserting control rods between the fuel rods. Control rods choke off the flow of neutrons and thus end the chain reaction, shutting down the reactor.

If the reaction cannot be stopped, the plant could experience a runaway reaction.

If engineers decide that the shroud must be repaired, divers could bolt braces into place, he said, although the job would be difficult.

At the Union of Concerned Scientists, a nuclear watchdog group, Robert Pollard, an engineer who was once an inspector for the Nuclear Regulatory Commission, said that such cracks were "one of the longstanding concerns about embrittlement of reactor internals."

Figure 0.25 Report on cracks in nuclear power plant.

the centerline. The area of the plane was near where the swing-wings were attached. Cracks were found in 38 of the 97 planes in the force.

Metal (steel, galvanized cast iron, copper, etc.) pipes and plastic (PVC, PE, etc.) pipes are also man-made structures that can lead to huge economic losses and human suffering if failure occurs prematurely. Stress corrosion cracking (SCC) in steel gas pipelines resulted in rapid crack propagation (RCP) and catastrophic final failure as shown in Figure 0.26 (a and b). To classify pipes

Figure 0.26 (a) Colony of stress corrosion cracks inside the steel pipes leading to dynamic fracture and (b) catastrophic failure in pipe.

Figure 0.27 Full-scale RCP testing.

that are susceptible to such catastrophic failure (by RCP), oil and natural gas companies developed a very expensive test procedure known as full scale testing to mimic actual failure. The test is supposed to closely resemble actual failure in scale and applied loading conditions. The test involves burying the pipes with a preexisting prescribed size defect followed by pressurization till catastrophic failure (Figure 0.27) occurs. It is clear from the picture that the scale of explosion of such failure is enormous. Apart from the large cost associated with these tests, there are significant difficulties conducting these tests. To establish statistical basis, many such tests needed to be run to establish the pipes constructed from certain material for field use. This makes the development of a better pipe qualified for field use almost impossible.

Figure 0.28 presents methods developed for the inspection of steel gas pipes for SCC. The process involves unearthing portions of the pipeline and arming it with ultra-scan sensors. Figure 0.29 shows the housing that carries several such sensors which finally get installed in the pipeline. One such installation operation is shown in Figure 0.30. Many efforts were made in universities and standard generating organizations like ISO and ASME to qualify a laboratory scale RCP test [Leevers et al. (1992)] to prequalify or screen a pipe/pipe material before subjecting to a full-Ssale test. Small scale steady state (S4) is such a test for RCP. Figure 0.31 shows an example of a pipe before and after such a test. The test was done under a test pressure of 4.98 bar on a plastic pipe (HDPE) of diameter 114.3 mm and initial crack length to diameter ratio of 5.9. The top view in Figure 0.31a shows the

Figure 0.28 Inspection operation in a portion of an excavated natural gas transmission pipeline for SCC.

*Figure 0.29 Thirty-six-*inch ultrasonic sensor carrier.

Figure 0.30 Launching the ultrasonic carrier in the excavated pipeline.

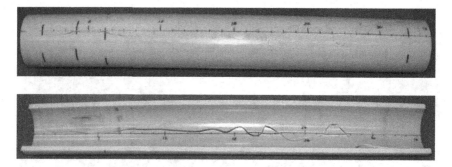

Figure 0.31 (a) Dynamic fracture (RCP) in plastic (HDPE) pipe during S4 testing and (b) inner wall of the fractured pipe showing the path of dynamic fracture.

properly marked pipe, and the bottom view shows the inner wall of the pipe after RCP was initiated by S4 test procedure.

Figure 0.32b shows dynamic fracture (RCP) in an eight-inch diameter PVC pipe after a field test. The path of crack propagation on the side wall is visible on the top view (Figure 0.32a). Copper tubing also highlights the region of crack discontinuities with rectangular boxes. It appears that a very early-stage bifurcation occurred. The upper crack unloaded due to shielding effect and got arrested [Ravi-Chandar and Knauss, 1984; Carlsson and Isaksson

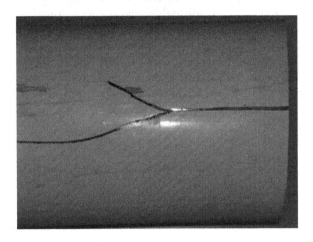

Figure 0.32 (a) Dynamic fracture (RCP) in plastic (PVC) pipe after field dynamic testing; (b) closer view where crack paths bifurcated where upper crack unloads as lower crack runs to final catastrophic failure.

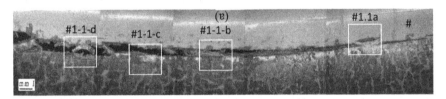

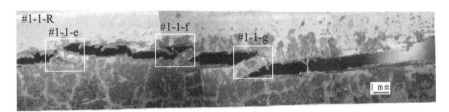

Figure 0.33 (a) Stress corrosion cracking in copper tubing with boxes around the crack discontinuities 1.1 a, b, c, and d; (b) continuation of the tube showing discontinuities 1.1 e, f, and g.

(2019)], while the lower crack continued to final failure as highlighted in the bottom view (Figure 0.32b).

Figure 0.33 shows a copper tubing that underwent SCC. The upper and lower figures present the stress corrosion cracks with several discontinuities as it progressed. Each discontinuity is marked by rectangular boxes. A long list of catastrophic failures has changed our trust in existing design. The main challenge is the determination of structural reliability (probability of "No Failure" during a time interval t). This topic involves invoking disciplines that are not addressed by existing materials science and mechanics community. This book intends to introduce these disciplines so that our future engineers get a chance to develop more reliable structures instead of depending on chances.

Before diving into the heart of this book, it is best to quickly review the existing strength of materials and mechanics with appropriate references in a more historical context so that we can appreciate the current weakness of this field.

Chapter 1

Review of Classical Strength of Materials

Sir Arthur Edington's (Sir Arthur Stanley Eddington was an English astronomer, physicist, and mathematician) advice to the enquirers was don't ask a physicist what "length" is?; just observe how one measures the length and make your definition accordingly. Thus, following his advice, let us observe how practical engineers measure strength. The first attempt to formulate strength criterion of materials is evidenced in Galileo Galilei's notebook where he proposed a new science, the study of the strength of materials, which considered how the size and shape of structural members affect their ability to carry and transmit loads (Figure 1.1). The maximal load the element can carry without failure may serve as the strength criterion. Much later, when the scientists had to introduce the stress tensor, Galileo's criterion was transformed into the maximal principal stress criterion. You cannot, therefore, simply double or triple the dimensions of a beam and expect it to carry double or triple the load. In modern notation, one can express this criterion as follows, where F_c is the critical load and σ_c is the critical stress at failure:

From F (Force) $= F_c$; to $F/A = \sigma_c$ stress criterion; A is the cross-sectional area on which F is applied

If one expresses F/A as stress and monitors change in length (displacement) ΔL, with a dimensionless quantity strain ε ($\Delta L/L_0$; where L_0 is the original length), it is possible to construct a diagram called engineering stress–strain relationship (shown in the diagram in Figure 1.2).

In materials that can undergo large deformation, elastic limit depicts the stress–strain limit under which the specimen demonstrates mechanically and thermodynamically reversible (no heat dissipation) response. In most materials, there is a very small portion within which this limit is linear. The slope of the curve for this linear portion is depicted as an initial modulus. Beyond elastic limit, strain softening leads to nonlinearity in stress–strain relation. Up to a certain strain level, the specimen may still be mechanically

DOI: 10.1201/9781003359845-2

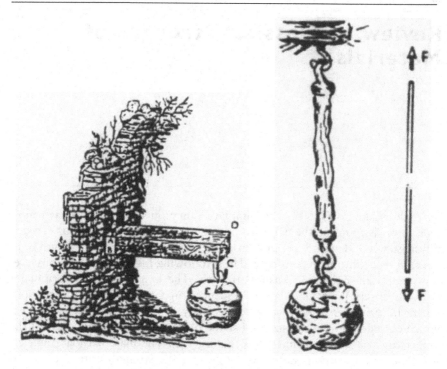

Figure 1.1 Schematic from Galileo Galilei's notebook.

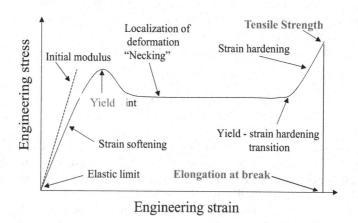

Figure 1.2 Schematic of engineering stress–strain diagram.

and thermodynamically reversible. On further displacement, thermodynamic irreversibility is noticed with associated heat dissipation. Ultimately softening leads to a yield point where specimen undergoes large displacement without any change in stress. Since this displacement at the yield point is relatively fast compared to the rate of change in applied displacement at the specimen boundary, an inflection point appears as shown in Figure 1.2. As one approaches the yield point, both mechanical and thermodynamic reversibility is lost, and specimen starts to show permanent deformation. In certain materials like plastics, this large displacement occurs instantaneously leading to neck formation as the first irreversible transition. Once the neck is formed, displacement occurs by transformation of original material at the boundary of the neck. This localized deformation at yield is coined as "necking". In metals, this necking phenomenon is often replaced by continuous deformation referred to as plastic deformation. Often during this process of large displacement, materials at micro or molecular scale undergo a second transition referred to as strain hardening ultimately leading to failure. This point of failure is characterized by tensile strength and elongation (or strain) to break. Stress–strain response leading to final failure as mentioned earlier is referred to as ductile. For many engineering applications, the yield point is the practical use limit and naturally attributed as strength of materials instead of final rupture strength. There are many conditions under which materials fail to undergo large deformation to yield, and such failure is referred to as brittle failure. Brittle failures do not go through strain softening, yield, and post yield. Another approach to dealing with strength of materials is to observe the failure types. *It would soon be clear that there is no brittle or ductile material, there is only brittle or ductile failure response.* These responses depend on loading history. Even a material that fails in brittle mode can fail in a ductile manner under suitable loading conditions.

Classification of Fracture Types

Figure 1.3 introduces the traditional classification of fracture/failure types associated with loading types, and on the right side of the failure type diagrams, the concept of stress as a tensor is introduced. Once it was recognized that any point in a three-dimensional object under external stress is experiencing stress in several directions, it was necessary to express stress as a second-rank *tensor* (see Chapter II) where the diagonal elements are referred to as normal stresses and rest as shear stresses. This led us to a vast field of stress analysis in mechanics of solids and subsequently the development of classical strength of materials. Stress and strain state conditions under which the material failure is observed are called "Strength Criteria". Figure 1.3 left column identifies load applied to various specimen geometries. The first two rows represent load applied along the center axis of a solid cylinder. The next two rows show torsional load applied to a hollow cylinder and bending load applied to a rectangular beam. On the application of such loading, two types of failure response may ensue depending on the internal structure of the material and external variables such as temperature and pressure. Column 2 shows the nature of brittle failure, and column 3 shows the nature of ductile failure on the application of the same tensile loading type. In brittle case, on application of tensile load, simple separation occurs perpendicular to the

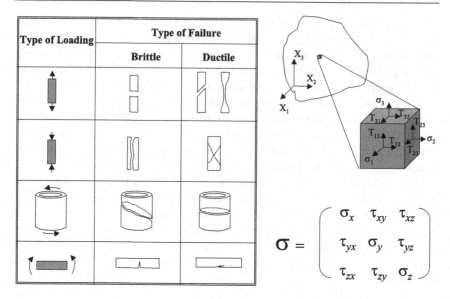

Figure 1.3 Types of failure associated with loading type and the introduction of stress tensor.

loading direction. In ductile cases, two types of failure can occur for the same tensile loading type. One is associated with slip band and shear band. One leads to neck formation or cold drawing (row 1). Row 2 shows failure types on application of compressive load. Brittle failure occurs by splitting of the cylinder into halves, whereas in ductile case, failure occurs by approximate 45° slip. Row 3 shows failure at 45° in brittle mode with clean fracture surface. Row 3 column 3 shows ductile splitting of the hollow cylinder. Row 4 columns 2 and 3 show brittle and ductile type failure. In brittle mode, a notch may appear from a defect site. In ductile mode, the notch appears causing slippage at the site.

Stress Analysis and Failure Criteria

Thus, one can simplify the classical strength of materials into two parallel developments: (a) Stress analysis and (b) material strength criteria as depicted in the diagram in Figure 1.4. Entropy criterion by Chudnovsky et al. is the latest of the criteria developments and will be introduced in detail in a later section. This is the only criterion that is capable of predicting time to failure where all other criteria fail. Today's field of stress analysis has matured to the state that one can solve any stress state for any type of object either analytically or numerically to a high degree of accuracy. Our designers are equipped with these results to apply to any complex structure an artist can imagine.

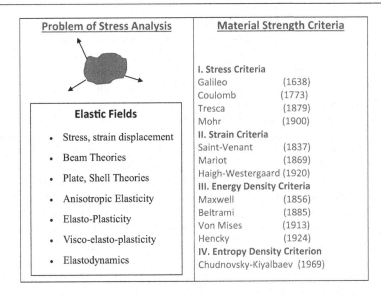

Problem of Stress Analysis	Material Strength Criteria

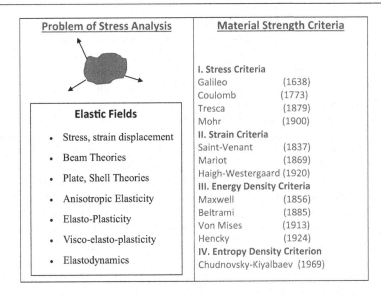

Elastic Fields

- Stress, strain displacement
- Beam Theories
- Plate, Shell Theories
- Anisotropic Elasticity
- Elasto-Plasticity
- Visco-elasto-plasticity
- Elastodynamics

I. Stress Criteria
Galileo   (1638)
Coulomb   (1773)
Tresca   (1879)
Mohr   (1900)

II. Strain Criteria
Saint-Venant   (1837)
Mariot   (1869)
Haigh-Westergaard (1920)

III. Energy Density Criteria
Maxwell   (1856)
Beltrami   (1885)
Von Mises   (1913)
Hencky   (1924)

IV. Entropy Density Criterion
Chudnovsky-Kiyalbaev (1969)

Figure 1.4 Problem of stress analysis and the evolution of strength criterion.

But how long these structures are going to last is still a mystery to them, and they tend to deal with it empirically.

This book intends to advance new concepts to remove this mystery. Considering left hand side of the table in Figure 1.4, we are going to summarize some of the key stress analysis techniques in the following paragraphs. *Stress, strain, and displacement* concepts on the left column of Figure 1.4 were already discussed. Stress–strain analysis (or stress analysis) is an engineering discipline that uses many methods to determine the stresses and strains in materials and structures subjected to forces. In continuum mechanics, stress is a physical quantity that expresses the internal forces that neighboring particles of a continuous material exert on each other, while strain is the measure of the deformation of the material. The purpose of stress analysis is to mathematically describe stress and strain (normalized displacement) fields for a given geometry of an elastic body. Material properties in continuum scale needed for failure of structures in elastic regime are initial modulus/Young's modulus/elastic modulus $(E = \sigma\varepsilon)$, where (σ) is stress and (ε) is the applied strain, yield strength, ultimate strength/rupture strength (defined earlier) and Poisson's ratio (υ). Poisson's ratio named after Siméon Poisson is the negative of the ratio of transverse strain to lateral or axial strain.

Stress analysis is a primary task for civil, mechanical, and aerospace engineers involved in the design of structures of all sizes, such as tunnels, bridges and dams, aircraft and rocket bodies, mechanical parts, and even plastic cutlery and staples. Stress analysis is also used in the maintenance of such structures and to investigate the causes of structural failures. Typically, the starting point for stress analysis is a geometrical description of the structure, the properties of the materials used for its parts, how the parts are joined, and the maximum or typical forces that are expected to be applied to the structure. The output data is typically a quantitative description of how the applied forces spread throughout the structure, resulting in stresses, strains, and the deflections of the entire structure and each component of that structure. The analysis may consider forces that vary with time, such as engine vibrations or the load of moving vehicles. In that case, the stresses and deformations will also be functions of time and space. In engineering, stress analysis is often a tool rather than a goal in itself; the ultimate goal being the design of structures and artifacts that can withstand a specified load, using the minimum amount of material. Stress analysis may be performed through classical mathematical techniques, analytic mathematical modelling or computational simulation, experimental testing, or a combination of methods.

Classical beam theory (second in the table of stress analysis) is a simplification of the linear theory of elasticity which provides a means of calculating the load-carrying and deflection characteristics of beams [Timoshenko, S. (1953), Timoshenko, S. and Woinowsky-Krieger, S. (1959), Timoshenko, S. (1976)]. It's expressed as the Euler–Bernoulli equation that describes the relationship between the beam's deflection and the applied load. In this equation, one more important term appears known as moment of inertia (I). I of a rigid body is a quantity that determines the torque needed for desired angular acceleration about a rotational axis. It is expressed in units of kilogram meter squared $\left(kg.m^2\right)$. It covers the case for small deflections of a beam, which are subjected to lateral loads only. It is thus a special case of generalized beam theory by Timoshenko et al. It was first enunciated circa 1750 but was not applied on a large scale until the development of the Eiffel Tower and the Ferris wheel in the late 19th century. Following these successful demonstrations, it quickly became a cornerstone of engineering during the second industrial revolution.

A plate and more generally a shell (*third in the table of stress analysis*) is a special three-dimensional body whose boundary surface has special features. In continuum mechanics, *plate* theories are mathematical descriptions of the mechanics of flat plates that draw on the theory of beams. Plates are defined as plane structural elements with a small thickness compared to the planar dimensions [Todhunter and Pearson (1886), Timoshenko and Woinowsky-Krieger (1959)]. The typical thickness to width ratio of a plate structure is less than 0.1. Plate theory takes advantage of this disparity in length scale to reduce the full three-dimensional solid mechanics problem to a two-dimensional problem. The aim of plate theory is to calculate the deformation and stresses in a plate subjected to loads. A shell is said to be *thin* if its thickness is much *smaller* than a certain characteristic length of the reference surface, e.g., the minimum radius of the curvature of the reference surface for initially curved shells.

Anisotropic elasticity, elastoplasticity, and viscoplasticity mentioned in the table of stress analysis (Figure 1.4) simply are enhancements of elasticity theory for special cases of unique material response to applied stresses. *Anisotropy* applies to elastic body where property varies in magnitude with the direction of measurement. *Elastoplasticity* applies to materials that display permanent deformation beyond a critical displacement or deflection level. Elastoplasticity is a rate-independent phenomenon which, when activated, causes the development of permanent deformation. It is usually considered as the limit case of viscoplasticity in which the notion of critical stress exists and causes yielding in the material. In metals, the development of plastic strains is due to the creation of dislocations in the crystalline structure. The evolution of the permanent deformation may or may not be linked to the criterion under which plasticity occurs, allowing us to categorize the elastoplastic mechanisms into associated and nonassociated. Viscoplasticity is a theory in continuum mechanics that captures the rate-dependent inelastic behavior of solids. Rate-dependence in this context means that the deformation of the material depends on the rate at which loads are applied. The inelastic behavior means irreversible deformation. Rate-dependent plasticity is important for transient plasticity calculations. The main difference between rate-independent plastic and viscoplastic material models is that the latter not only exhibit permanent deformations after the application of loads but also continue to undergo a creep flow as a function of time under the influence of the applied load [Perzyna, P. (1966)]. The elastic response of viscoplastic materials can be represented in one dimension by Hookean spring elements. Rate-dependence can be represented by nonlinear dashpot elements in a manner like viscoelasticity. Plasticity can be accounted for by adding sliding frictional elements [Lemaitre and Chaboche (2002)].

On the right column are the key failure criteria and their evolution. We already talked about Galileo's critical stress criteria (strength). Coulomb stress criteria appeared about 100 years after Galileo in an essay around 1773. It was not till 1900 that this proposition was further expanded by Mohr and became popular as Mohr–Coulomb stress criteria [Timoshenko, S. (1976)]. Generally, the theory applies to materials for which the compressive strength far exceeds tensile strength. The Mohr–Coulomb failure criterion represents the linear envelope that is obtained from a plot of the shear strength of a material versus the applied normal stress. This relation is expressed as

$$\tau = \sigma \tan \varphi + C \tag{1.1}$$

where τ is the shear strength, σ is the normal stress, C is the intercept of the failure envelope with the τ axis, and $\tan \varphi$ is the slope of the failure envelope. If compression is assumed to be negative, then σ should be replaced with $-\sigma$. If $\varphi = 0$, the Mohr–Coulomb criterion reduces to the Tresca criterion. Tresca criterion (1879) is equivalent to saying that yielding will occur at a critical value of the maximum shear stress. Von Mises yield criterion considers that yielding of a ductile material begins when the second invariant of deviatoric stress reaches a critical value [Von Mises, R. (1913); Hill, R. (1950)]. It is a part of plasticity theory that mostly applies to ductile materials, such as some metals. Prior to yield, material response can be assumed to be of a nonlinear elastic, viscoelastic, or linear elastic behavior. In materials science and engineering, Von Mises yield criterion is also formulated in terms of the Von Mises stress or equivalent tensile stress. In this case, a material is said to start yielding when the Von Mises stress reaches a value known as yield strength, σ_y. The Von Mises stress is used to predict yielding of materials under complex loading from the results of uniaxial tensile tests. The Von Mises stress satisfies the property where two stress states with equal distortion energy have an equal Von Mises stress. Heinrich Hencky formulated the same criterion as Von Mises independently in 1924.

At about the same period, engineers observed that the presence of discontinuities in a structure could lead to the concentration of stress. A stress concentration (often called stress raisers or stress risers) is a location in an object where stress is concentrated. Figure 1.5 shows stress distribution in an

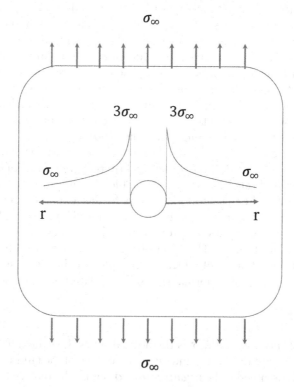

Figure 1.5 Stress distribution in an infinite plate on the application of remote force σ_∞.

infinite plate with a circular hole. It is clear from Figure 1.5 that *stress* tends to *concentrate* at the circumference of the hole $(\sigma_{max} = 3\sigma_\infty)$ perpendicular to the direction of applied force σ_∞ and then decays to stress at infinity. The story of the analysis of stress concentrations begins in 1898 with Ernst Gustav Kirsch's linear elastic solution for stresses around a hole in an infinite plate [Kirsch, E.G. (1898)]. Kirsch's solution contains the well-known factor-of-three stress concentration at the hole under uniaxial loading. But there is more to the story. The stress concentration can, in fact, vary from two to four for more complex loading conditions, i.e., stress states different from uniaxial tension.

In this book, we will only review Kirsch's solution for the basic case of uniaxial tension which can be extended to more complex cases of equibiaxial tension and shear using superposition principles. Kirsch's solution for stresses at a hole is for the case of uniaxial tension in an infinite plate. Uniaxial tension is represented by the remote stress, σ_∞. The hole has radius, a, the radial coordinate is r (except for $r < a$), and $\theta = 0$ aligns with the remote loading direction. We will see that the famous factor-of-three stress concentration occurs at $\theta = \pm 90°$. The solution for the stress state around a hole is

$$\sigma_{rr} = \frac{\sigma_\infty}{2}\left(1-\left(\frac{a}{r}\right)^2\right)+\frac{\sigma_\infty}{2}\left(1-4\left(\frac{a}{r}\right)^2+3\left(\frac{a}{r}\right)^4\right)cos2\theta \qquad (1.2)$$

$$\sigma_{\theta\theta} = \frac{\sigma_\infty}{2}\left(1+\left(\frac{a}{r}\right)^2\right)-\frac{\sigma_\infty}{2}\left(1+3\left(\frac{a}{r}\right)^4\right)cos2\theta \qquad (1.3)$$

$$\tau_{r\theta} = -\frac{\sigma_\infty}{2}\left(1+2\left(\frac{a}{r}\right)^2-3\left(\frac{a}{r}\right)^4\right)sin2\theta \qquad (1.4)$$

At, $r = \infty$, all the a/r terms go to zero, leaving

$$\sigma_{rr} = \frac{\sigma_\infty}{2}+\frac{\sigma_\infty}{2}cos2\theta \qquad (1.5)$$

$$\sigma_{\theta\theta} = \frac{\sigma_\infty}{2}-\frac{\sigma_\infty}{2}cos2\theta \qquad (1.6)$$

$$\tau_{r\theta} = -\frac{\sigma_\infty}{2}sin2\theta \qquad (1.7)$$

and $\sigma_{rr} = \sigma_\infty$ when $\theta = 0°$ and $180°$, while $\sigma_{\theta\theta} = \sigma_\infty$ when $\theta = \pm 90°$. The shear stress, $\tau_{r\theta}$, is simply the result of coordinate transformations on σ_∞.

At the hole, at $r = a$, all the a/r terms equal 1, producing

$$\sigma_{rr} = 0, \sigma_{\theta\theta} = \sigma_\infty\left(1-2cos2\theta\right) \qquad (1.8)$$

$$\tau_{r\theta} = 0 \qquad (1.9)$$

The radial stress, σ_{rr}, and the shear stress, $\tau_{r\theta}$, are 0 at the hole because it is a free surface. Hoop stress, $\sigma_{\theta\theta}$, merits attention. At $\theta = 0$, $\sigma_{\theta\theta} = -\sigma_\infty$, so the hoop stress is compressive. However, it is at $\theta = \pm90°$ that $\sigma_{\theta\theta} = 3\sigma_\infty$ and the factor-of-three stress ratio occurs. This ratio is called the *Stress Concentration Factor* (K) and is discussed later in the chapter. Note that the stress components at the hole are independent of the size of the hole itself. This is related to the fact that the plate is infinitely large, so the hole's size is inconsequential relative to the plate. The *Stress Concentration Factor, K*, is the ratio of maximum stress at a hole, fillet, or notch (but not a crack) to the remote stress. For our case of a hole in an infinite plate, $K = 3$. Following the exact theory of Kirsch, engineers started developing solutions for K for plates with finite boundaries, semi-circular edge grooves, and fillets considering geometric details. Figures 1.6 and 1.7 show examples of such solutions [Timoshenko, S. (1976)]. We must not confuse the Stress Concentration Factor here with the *Stress Intensity Factor* (SIF), K_I, used in crack analyses. The two are completely different. For starters, a Stress Concentration Factor is dimensionless, while an SIF is not. SIF will be discussed in detail on later pages.

In the mid-1900s, scientists and engineers focused their energy on plotting such diagrams since they realized the structures could fail due to these concentrated stresses. It was quickly recognized that a body is not just subjected to these designed stress concentration sites but many visually unobservable defects/flaws within it of much smaller dimension. C.E. Inglis is the first person to take a major step in the direction of quantification of the effects of crack-like defects. He started his stress analysis starting with an elliptical hole, but as the major axis, a, increases relative to the minor axis, the hole begins to take the shape of a sharp crack-like defect. If ρ is the radius of

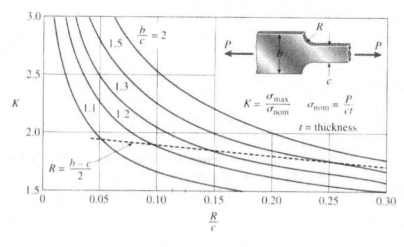

Figure 1.6 Stress Concentration Factor K for flat bars with shoulder fillets.

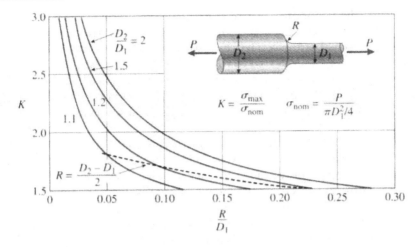

Figure 1.7 K for round bars with shoulder fillets.

curvature at the ends of the major axis of the ellipse, it would yield the following famous equation for crack tip stress where σ_∞ is the stress applied at infinity [Inglis, C.E. (1913)]. Inglis' solution is a simple extension of Kirsch's solution of Stress Concentration Factor.

$$\sigma_t = \sigma_\infty \left(1 + 2\sqrt{\frac{a}{\rho}}\right) \tag{1.10}$$

So, attention was focused on crack tip stress, and a stress-based criterion was proposed where crack tip stress reaches a point to break bonds. Another such criterion was proposed by Westergaard. In 1939, Harold M. Westergaard developed a solution for the stress field surrounding a crack [Westergaard, H.M. (1939)] that has one big advantage over Inglis' solution as Westergaard's function (Z_I) applies directly to cracks, not to an ellipse that approaches a crack in an infinite plate at its limit. Z_I is a complex stress function. Z_I was later related to SIF K_I for uniaxial loading as follows where ℓ is the crack length [Parker, A.P. (1981), Anderson, T.L. (1991)]:

$$K_I = (2\pi)^{1/2} \lim_{z \to \ell} (z - \ell)^{1/2} Z_I \tag{1.11}$$

In its original form, this function is suitable for solving a limited range of crack problems. It leads to the expected inverse square root stress singularity expected from the solutions of cracked bodies utilizing theory of elasticity discussed later.

Now let's carefully consider right hand side of the table listing all classes of strength criteria. Galileo was the first to introduce the stress criterion after recognizing the importance of stress. In later years, recognizing that stress is a tensor, several criteria were proposed by Coulomb, Tresca, and Mohr involving principal stresses, deviatoric stresses, and stress invariants (Von Mises criterion) in later years [Von Mises, R. (1913), Tresca, H.S. (1864), Hill, R. (1950)]. Once the stress-based criteria were exhausted, strain criteria started making their mark as shown in Figure 1.4. It was not too difficult to extend stress–strain to some energy criteria by several scientists in mid-1800 to 1900s.

As mentioned earlier, entropy density criterion made its mark in 1969. This criterion was published in a Russian journal that did not receive attention from regular people in this field. Furthermore, researchers in this field lack foundation in irreversible thermodynamics [Kiyalbaev, D.A. and Chudnovsky, A.I. (1969)] resulting in limited acceptance. Sadly, the first three types of criteria mentioned in Figure 1.4 continued to be used till modern times in spite of their limited applicability to only materials under specific external load condition. It is well known that when scientific theories grow at a fast rate to explain a single phenomenon without true success, it is imperative that everyone has missed the true understanding. As we will show that entropy density criterion allows one to address time to failure as a part of failure criterion clearly absent earlier.

Next few paragraphs are dedicated to a brief exposition of the *Entropy Density Criterion of long-term strength of the engineering materials*. The entropy criterion was proposed by A. I. Chudnovsky and elaborated in a joint paper [Kiyalbaev, D.A. and Chudnovsky, A.I. (1969)]. Almost eight years later, chapter 7 of a book on the strength of materials in heavy machine building by three authors was also dedicated to a review of the entropy density criterion [Goldenblat et al. (1977)]. The authors claimed that they came up with the entropy density criterion later but independently in page 207 of the second paragraph. Incidentally, the first author of the book possibly had a chance to get familiar with the proposal on entropy density criterion by being appointed by the supreme qualification committee as an undercover reviewer of Chudnovsky's doctorate thesis. Chudnovsky thesis was later published in the book [Chudnovsky, A.I. (1973)]. The entropy density criterion has several advantages in comparison with the classical strength criteria such as the maximum principal stress criterion; the first strength theory and the maximum shear stress criterion; and the second strength theory. The classical strength theories do not address the issue of time to the first local failure, which is a crack-initiation time in contrast with the entropy density criterion leading to an explicit expression for the time of the local failure [Goldenblat et al. (1977)].

$$\frac{T\Delta S^*}{B\sigma_\theta^{m+1}} \qquad\qquad (1.12)$$

where T is the absolute temperature, ΔS^* is critical entropy increment at which local failure occurs (must be determined experimentally), σ_θ is the shear stress intensity, and B and m are the creep coefficient of creep and exponent, respectively, based on the creep equation of power law type. Entropy density criterion stems from analogy between metallic crystal failure and melting/sublimation [Born (1939, 1940)]. This makes possible the following assumption that certain critical value of entropy density S^* is a property of material, corresponds to a local failure, and is expressed as $s(t^*) = s(0) + \Delta S^*$. Here, t^* is the instance of failure occurrence and $s(0)$ is the entropy density at the initial instant of time. Since all these phenomena appear from atomic bond instability, the application of analogy makes sense. It suggested critical energy-based criterion like latent energy. However, it was known that latent energy of melting and sublimation depends on pressure and temperature. Similarly, stress (σ) dependency is critical for local failure as pressure is related to stress (σ). At the same time, the ratio of latent energy to temperature (i.e., entropy) turns to be a material constant [Swalin, R. (1972)]. In addition, the aforementioned entropy density criterion leads to the linear summation of the damage accumulation events, independent of the source. This criterion also opens a way to directly account for the contributions of various physico-chemical processes to local fracture such as diffusion, chemical reactions, and corrosion, since all these processes contribute to the internal entropy production [Kiyalbaev, D.A. and Chudnovsky, A.I. (1969)]. Thus, entropy density criterion allows to model mechano-chemical effects in fracture. At this point, we would like to introduce a few paragraphs on the utilization of this knowledge in actual field failures.

Lessons from Case Studies

Here, we summarize Dr. A. Chudnovsky's contribution as expert witness in civil litigation of premature failure of PB tubing in water distribution systems. Traditionally, copper tubing was commonly used for hot- and cold-water distribution systems in single homes as well as multi-floor buildings. However, at the end of 1970, the price of the copper went up, which made the usage of copper too expensive, and Shell Chemical company developed a new product such as polybutylene (PB) tubing with polyacetal connectors to substitute the expensive copper system. The developer provided a 50-year warranty for the product without proper testing. That was a fatal mistake. The PB tubing system was widely advertised and became a commercially successful product that has been widely used in various states of the United States such as California, Texas, Alabama, Georgia, Pennsylvania Maryland, Ohio as well as in Canada; unfortunately, the PB water distribution system started to fail much earlier than the life span of the system, promised by the warranty that triggered a number of litigations in the named aforementioned states. Dr. A. Chudnovsky (AC) was asked to serve as an expert witness on the plaintiff's side in the number of such cases. He determined that the root cause of the premature failure was the so-called mechano-chemical effect, well-known in metallic materials as the stress corrosion. However, the developer of the PB water distribution system claimed that the PB in contrast with metals does not corrode, and thus they promised 50 years of service life without testing the system, since the testing for lifetime is a long and expensive proposition. The failure of the PB systems typically started from the weakest link in the system: polyacetal fittings. In the forensic studies of the acetal fittings failure, AC had a very good and highly qualified partner: Dr. Salvatore Stivala. Dr. Stivala was an organic chemist well known for his books on polymer degradation, specifically oxidative degradation of polymers. The oxidative degradation of the polyacetal in combination with the mechanical stress in polymers is the direct analogy of the stress corrosion cracking (SCCC) in metals. Dr. A. Chudnovsky and Dr. S. Stivala jointly served as expert witnesses' team and testified in more than ten litigations in various states of the United States. A couple of decades later, AC worked on a project sponsored by the Gas Research Institute (GRI) on stress-corrosion failure of the steel high-pressure natural gas transmission pipes. Finally, Attorney George Fleming started a Class Action suit on behalf of all the people who suffered financial losses caused by the significant reduction of price of properties due to the widespread failure of PB tubing systems. After a 3-month mediation in California, the case finally settled for about $2 billion.

Later, the same phenomena were observed by AC in the copper tubing failure in a new building at the Millennia Park district of Chicago, which contained many miles of copper tubing in water distribution. The last pressure test prior to completion revealed multiple cracking in the tubing; when the cracked tubing was transported to the Fracture mechanics and Materials Durability Lab at the UIC for analysis, Dr. Chudnovsky diagnosed the stress corrosion cracking as the direct cause of failure. That was a puzzling diagnosis for AC himself since stress corrosion requires a combination of two major factors: (a) Mechanical stress and (b) a chemically aggressive environment. The brand-new tubing that had not been in service yet had no such factors. An intriguing story of the search for the origin of the SCC is described in one of the chapters of this book. All the practical cases with necessary theoretical and experimental details will be discussed in this book. Forensic studies of the premature failure of the engineering structures have motivated AC's interest in accelerated testing and addressing Longevity and Reliability of Engineering Structures, where Reliability means Probability of no Failure prior to the specified Time. Many years of research in that area is reflected in this book.

Chapter 2

Commonly Used Mathematical Tools

At this point, we need to introduce mathematical tools needed for analysis. In primitive societies of hunters and gatherers, there was a dependency on land area. They needed to learn how to measure and that was the beginning of Geodesics (study of distances on curved surface) and Geometry (study of distances on flat plane). If we position ourselves at any point on earth's surface on northern hemisphere, we can notice three independent directions (North, East, and West). Thus, any motion (characterized by velocity) on earth has direction. To characterize velocity, we needed to introduce the concept of vector which has both magnitude and direction (as shown in Figure 2.1 by x, y, and z coordinates). In contrast, scalar quantities only have magnitude. Examples of scalar quantities include mass, speed, distance, time, volume, density, and temperature. Examples of vector quantities include linear velocity and momentum, acceleration, displacement, angular velocity, force, electric field, and polarization.

We can then introduce summation and multiplication of vectors. When we multiply vectors, there are two types of multiplications such as dot product and cross product. Dot product establishes the correspondence between two vectors and a scalar. It is a product (also known as inner product), an operation that takes two vectors (e.g., $\vec{a}$ and $\vec{b}$) and returns a scalar quantity ($|a||b|\cos\theta$ where θ is the angle between the two vectors $\vec{a}$ and $\vec{b}$) as a result of the multiplication. If any two given vectors are said to be orthogonal, i.e., the angle between them is 90, then $\vec{a}.\vec{b} = 0$ as cos 90 is 0. If the two vectors are parallel to each other, $\vec{a}.\vec{b} = |a||b|$ as cos 0 is 1. The dot product fulfills the following properties if $\vec{a}$, $\vec{b}$, and $\vec{c}$ are real vectors and r is a scalar.

1. Commutative:

$$\vec{a}.\vec{b} = \vec{b}.\vec{a}$$

2. Distributive over vector addition:

$$\vec{a}.\left(\vec{b} + \vec{c}\right) = \vec{a}.\vec{b} + \vec{a}.\vec{c}$$

DOI: 10.1201/9781003359845-3

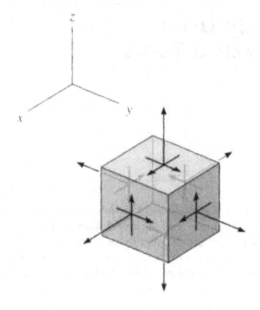

Figure 2.1 Directionality of a vector is shown by x, y, and z coordinates.

3. Bilinear:

$$\vec{a}.\left(r\vec{b}+\vec{c}\right)=r\left(\vec{a}.\vec{b}\right)+\left(\vec{a}.\vec{c}\right)$$

4. Scalar multiplication:

$$\left(c_1\,\vec{a}\right).\left(c_2\,\vec{b}\right)=c_1c_2\left(\vec{a}.\vec{b}\right)$$

5. *Not associative* because the dot product between a scalar $(a.b)$ and a vector $\vec{c}$ is not defined, which means that the expressions involved in the associative property are both ill-defined.

6. Orthogonal:

$$\vec{a}.\vec{b}=0$$

7. No cancellation: Unlike the multiplication of ordinary numbers, where if $a.b = a.c$, then b always equals c unless a is zero, the dot product does not obey the cancellation law.

8. Product rule: If a and b are (vector-valued) differentiable functions, then the derivative (denoted by a prime ') of $a.b$ is given by the rule $(a+b)' = a'.b + a.b'$.

Before 1844, mathematicians used to talk about a cross product that establishes correspondence between two vectors and a new object which is called second-rank asymmetric tensor. In 1844, a mathematician named Hermann

Grassmann (Grassmann Algebra) introduced an external product or bivector. This product leads to second rank asymmetric tensor. It took 44 years before it was finally introduced in western mathematics by English Mathematician William Kingdon Clifford. Sometimes Grassmann algebra is called Clifford algebra; unfortunately, this type of mistake has happened many times in the history of science. Another approach is to use a terminology called rank of a tensor for generalization. The total number of indices required to identify each component uniquely is equal to the dimension of the array, and is called the order, degree, or rank of the tensor. A first-rank tensor is a vector, a one-dimensional array of numbers. A second rank tensor is represented by a typical square matrix. Stress, strain, thermal conductivity, magnetic susceptibility, and electrical permittivity are all second-rank tensors. A simple example of a geophysical relevant tensor is stress. Stress, like pressure, is defined as force per unit area. Pressure is isotropic, but if a material has finite strength, it can support different forces applied in different directions. Figure 2.2 illustrates a unit cube of material with forces acting on it in three dimensions. By dividing by the surface area over which the forces are acting, the stresses on the cube can be obtained. Any arbitrary stress state can be decomposed into nine components (labeled σ_{ij}). These components form a second-rank tensor—the stress tensor (Figure 2.2). In mathematics, the cross product or vector product is a binary operation on two vectors in a three-dimensional-oriented Euclidean vector space and is denoted by the symbol $\otimes$. Given two linearly independent vectors a and b, the cross product is a $\otimes$ b (read "a cross b"). If two vectors have the same direction or have the exact opposite direction from each other (i.e., they are not linearly independent), or if either one has zero length, then their cross product is zero. The cross product is anticommutative (i.e., a $\otimes$ b $= -$b $\otimes$ a) and is distributive over addition (i.e., a a $\otimes (b + c) =$ a $\otimes$ b $+$ a $\otimes$ c).

Another important parameter needed to deal with tensorial algebra is Kronecker delta (δ_{ij}). Kronecker delta, named after Leopold Kronecker, is a function of two variables (nonnegative integers). The function is 1 if the variables are equal, and 0 otherwise. Thus, Kronecker delta is essentially the dot product of two orthogonal unit basic vectors.

$$\delta_{ij} = \begin{cases} 0 \ if \ i \neq j \\ 1 \ if \ i = j \end{cases} \qquad (2.1)$$

Kronecker delta is a very useful symbol in mathematics, physics, and engineering because of its shifting property. Inner (scalar) product of vectors can be written as

$$\vec{a} \cdot \vec{b} = \sum_{i,j}^{n} a_i \delta_{ij} b_j = \sum_{i}^{n} a_i b_i \quad for \ i = j \qquad (2.2)$$

Tensorial ranks or rank of a tensor T is the minimum number of simple tensors that sum up to T. The rank of a tensor is the number of indices. The first

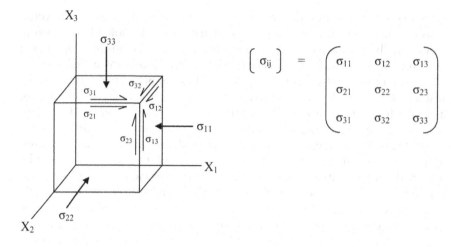

Figure 2.2 Forces acting on three dimensions leading to a stress tensor.

three ranks (also called orders) for tensors (0, 1, 2) are scalar, vector, and matrix. Although these three are technically simple tensors, a mathematical object isn't usually called a "tensor" unless the rank is 3 or above.

Nabla Operator is another operator used in mathematics as a vector differential operator, usually represented by a symbol ∇, and, when applied to a scalar function such as density or temperature, it provides the gradient (it denotes the standard derivative of the function as defined in calculus). When applied to a field (a function defined on a multidimensional domain), it may denote any one of three operators depending on the way it is applied: The gradient or (locally) steepest slope of a scalar field (or sometimes of a vector field, as in the Navier–Stokes equations); the divergence of a vector field; or the curl (rotation) of a vector field. These three uses are summarized as:

Gradient: grad $f = \nabla f$
Divergence: div $\bar{v} = \nabla \cdot \bar{v}$
Curl: curl $\bar{v} = \nabla \otimes \bar{v}$

It's a vector that has magnitude and direction where the components are derivatives as in calculus. ∇, when applied to a vector field such as flow of energy (heat), provides divergence of the field. ∇, when applied via external product to a vector function, raises the tensorial rank, for example, stress/strain tensor (rank 2 tensor) from displacement vector (rank 1 tensor). Symmetry and anti-symmetry are linked to properties of a tensor. A tensor a_{ij} is symmetric if $a_{ij} = a_{ji}$ while a tensor b_{ij} is antisymmetric if $b_{ij} = -b_{ji}$. It follows that all diagonal components of an antisymmetric tensor are zero. Therefore, the antisymmetric tensor has only three independent components similar to a vector. That was the source of the confusion mentioned earlier. The inner product of a symmetric and antisymmetric tensor is also zero.

$$a_{ij}.b_{ij} = -a_{ji}.b_{ji}. \text{ Thus } 2a_{ij}.b_{ij} = 0 \qquad (2.3)$$

In a system which consists of unequal principal stresses, there are three deviatoric stresses, obtained by subtracting the mean (or hydrostatic) stress (p) from each principal stress (i.e., $\sigma_1 - p$, $\sigma_2 - p$ ˙, and $\sigma_3 - p$). *Deviatoric stresses* control the degree of body distortion, i.e., the shape. *Hydrostatic* or *dilatational* stress acts to change the volume of the material only. One more thing to introduce is the invariants of the stress tensor. Stress tensor is a second-rank tensor. The tensor consists of nine components σ_{ij} that completely define the state of stress at a point inside a material in the deformed state, placement, or configuration. This is conventionally expressed in a matrix form as shown next:

$$\sigma_{ij} = \begin{pmatrix} \sigma_{xx} & \tau_{xy} & \tau_{xz} \\ \tau_{yx} & \sigma_{yy} & \tau_{yz} \\ \tau_{zx} & \tau_{zy} & \sigma_{zz} \end{pmatrix}$$

where σ components represent principal normal stresses, and τ components represent shear stresses.

Stress invariants are represented by I_1, I_2, and I_3. These invariants can be expressed as $I_1 = \sigma_{xx} + \sigma_{yy} + \sigma_{zz}$, $I_2 = \sigma_{xx}\sigma_{yy} + \sigma_{yy}\sigma_{zz} + \sigma_{xx}\sigma_{zz} - \tau_{xy}^2 - \tau_{yz}^2 - \tau_{zx}^2$, and finally $I_3 = \det(\sigma_{ij})$. In a symmetric tensor, $\tau_{xy} = \tau_{yx}$ and so on. Invariants are very important properties of tensors since they remain unchanged if the coordinate system undergoes transformation such as rotation. There are certain invariants associated with the stress tensor, whose values do not depend upon the coordinate system chosen or the area element upon which the stress tensor operates. These are the three eigenvalues of the stress tensor, which are called the principal stresses.

Another mathematical tool that's exploited in this field is differential calculus. It's a subfield of calculus, which studies the rates at which quantities change. The primary objects of this subfield are dealing with derivative of a function, related notions such as the differential, and their applications. The derivative at a point is the slope of the tangent line to the graph of the function at that point, provided a derivative exists and can be defined at that point. The process of finding a derivative is called differentiation. Differential calculus is connected to the integral calculus since it is the reverse process of integration. An example of differentiation is when we determine the derivative of the displacement with respect to time returning as velocity (V). The derivative of the velocity with to respect time is acceleration (a).

The derivative of the momentum of a body of mass $m(mV)$ with respect to time equals the Newtonian force (F) applied on the body. Rearranging the derivative leads to the famous Newton's law $(F = ma)$ [*Appendix IV*]. Derivatives are frequently used to find the maxima and minima of a function. Equations involving derivatives are called differential equations. An ordinary differential equation (ODE) is an equation which consists of one or more functions of one independent variable along with their derivatives. A differential equation is an equation that contains a function with one or more derivatives. Only in the case of ODE, the word ordinary is used for the derivative of the functions for the single independent variable. The order of a differential equation is defined to be that of the highest-order derivative it contains. In general form, an nth-order differential equation can be expressed as $\frac{d^n y}{dx^n} + Py\frac{d^{n-1}y}{dx^{n-1}} + Qy^2\frac{d^{n-2}y}{dx^{n-2}} + \cdots Ky^n = I$. There is one more type of differential equation associated with multivariable functions. A partial differential equation is an equation which imposes relations between the various partial derivatives of a multivariable function.

As mentioned in the previous paragraph, the fundamental theorem of integral calculus connects the derivative and the integral, and it's the most common way to evaluate definite integrals. In a nutshell, it states that every continuous function over an interval has an antiderivative (a function whose rate of change, or derivative, equals the function). Furthermore, the difference $F(b) - F(a)$, where F is the function's antiderivative, is the definite integral of such a function over an interval $a < x < b$. Integral calculus is divided into two categories. They are described in subsequent sections. A function that takes the antiderivative of another function is called an indefinite integral. The indefinite integral is not defined with the help of upper and lower limits. Indefinite integral represents the family of the function whose derivatives are f. The difference between any two functions within the family is a constant. A definite integral has upper and lower limits (i.e., a start and an end value). x is limited to lying on a real line. When restricted to lying on the real line, a definite integral is also known as a Riemann Integral. For a discontinuous function, the integral is replaced by summation (Σ).

Another important operator is Laplace operator with a symbol Δ, which is a scalar operator that can be applied to either vector or scalar field. It is expressed as a dot product of two nabla operators $(\nabla \cdot \nabla = \nabla^2)$. The Laplace operator is named after the French mathematician Pierre-Simon de Laplace (1749–1827), who first applied the operator to the study of celestial mechanics: The Laplacian of the gravitational potential due to a given mass density distribution is a constant multiple of that density distribution. The Laplacian occurs in many differential equations describing physical phenomena.

Poisson's equation describes electric and gravitational potentials; the diffusion equation describes heat and fluid flow; the wave equation describes wave propagation and the Schrödinger equation in quantum mechanics. In rectangular coordinates, Laplacian can be expressed as a partial differential

of the form $\left(\dfrac{\partial^2 f}{\partial x^2} + \dfrac{\partial^2 f}{\partial y^2} + \dfrac{\partial^2 f}{\partial z^2} \right)$.

Chapter 3

Equations of Elasticity

All structural materials possess, to a certain extent, the property of elasticity, i.e., if external forces, producing deformation of a structure, do not exceed a certain limit, the deformation disappears with the removal of the forces. Most of the stress analysis techniques were developed for use of materials in structures that operate within these limits. There is a thermodynamic consideration that is omitted in elasticity that requires monitoring heat dissipated during the application and removal of forces. A true elastic material does not dissipate heat, and if one observes its loading and unloading curve, they will notice that both curves follow the same path. In many materials, there is a region where stress linearly rises with strain and the field is known as linear elasticity, and, beyond the linear elastic region, there would be a nonlinear elastic region. Within elastic regime, thermodynamic reversibility applies. Under such condition, elasticity equations are applicable.

There are three basic considerations in elasticity:

1. Elastic equilibrium: Two conditions of equilibrium must be satisfied to ensure that an object remains in static equilibrium. First, the net force acting upon the object must be zero ($\sum F_i = 0$, where F stands for forces), and momentum within the volume element sums up to zero ($\sum M_i = 0$, where M stands for momentum) such that the net torque acting upon the object will be zero.
2. Strain–displacement: Displacement refers to the change in position of a particle. When the displacement is very small, it is referred to as infinitesimal; when large, it is referred to as finite. Strain is also related to the displacement of particles from their original position to a new position. Strain and displacement are thus closely related. There is, however, a significant difference. Strain always involves changes in the internal configuration of the body. During strain, the distances between the particles that constitute the strained body change.
3. Stress–strain relations (constitutive relations)

DOI: 10.1201/9781003359845-4

Let us now consider several foundational equations of elasticity. Equilibrium equation is expressed as (implies no inner sources of stress):

$$\nabla \cdot \overset{\vee}{\sigma} = 0 \tag{3.1}$$

$$\partial_x \sigma_{xx} + \partial_y \sigma_{xy} = 0 \tag{3.2}$$

$$\partial_x \sigma_{yx} + \partial_y \sigma_{yy} = 0 \tag{3.3}$$

Hooke's Law provides the constitutive relations leading to following three equations:

$$\varepsilon_{xx} = \frac{1}{E}\left(\sigma_{xx} - \vartheta\sigma_{yy}\right) \tag{3.4}$$

$$\varepsilon_{yy} = \frac{1}{E}\left(\sigma_{yy} - \vartheta\sigma_{xx}\right) \tag{3.5}$$

$$\varepsilon_{xy} = \frac{1+\vartheta}{E}\left(\sigma_{yy}\right) \tag{3.6}$$

where ϑ is the Poisson's ratio and E is the Young's modulus or Elastic modulus. It is often useful to consider the relationship between stress and strain (opposite way). For this, we use compliance ($\varepsilon_{xx} = S_{xxyy}\left(\sigma_{yy}\right)$ where S_{xxyy} = compliance tensor). Young's modulus of elasticity in tension or compression (i.e., negative tension) is a mechanical property that measures the tensile or compressive stiffness of a solid material when the force is applied lengthwise. (Although Young's modulus is named after the 19th-century British scientist Thomas Young, the concept was developed in 1727 by Leonhard Euler. The first experiments that used the concept of Young's modulus in its current form were performed by the Italian scientist Giordano Riccati in 1782, pre-dating Young's work by 25 years.) λ and μ are also elastic properties and called Lamé moduli. They are related to the parameters E and ϑ, respectively the material Young's modulus and Poisson's ratio, by

$$E = \frac{\mu(2\mu+3\lambda)}{\mu+\lambda} \text{ and } \vartheta = \frac{\lambda}{2(\mu+\lambda)}.$$

Equilibrium equation expressed as the divergency of the stress tensor represented by letter σ equals zero means there are no sources of the inner forces or no body forces. The strain tensor is introduced as the symmetrical part of the gradient of the displacement vector field $u(x,y)$, and the antisymmetrical part of the same gradient represents the local rotation of the elementary domain in the process of deformation ($\tilde{\varepsilon} = \left(\overline{\nabla}\overline{u} + \overline{u}\overline{\nabla}\right)$). Prior to the understanding of this fact, the mathematicians erroneously used the cross product of the vector differential operator ∇ and $u(x)$.

The three components of strain are derived from the displacement vector field so that some restraints must be placed on "allowable strain". The strains must be compatible. This requirement yields the following compatibility equation.

$$\partial^2_{xx}\varepsilon_{yy} + \partial^2_{yy}\varepsilon_{xx} - 2\partial^2_{xy}\varepsilon_{xy} = 0 \tag{3.7}$$

Equations of equilibrium Eqs. (1.5) and (1.6) can be expressed in general form as

$$\partial_j\left(\sigma_{ij}\right) = 0 \tag{3.8}$$

The compatibility equations hold in an infinite plane when the Gaussian curvature R_{ijkl} is zero as shown later in the chapter. Physical meaning of such a claim is that during the application of strain, the intrinsic geometry does not change. If it was Euclidean, it remains Euclidean. R_{ijkl} of a surface at a point is the product of the principal curvatures. A cylinder is a surface of zero Gaussian curvature while a sphere has a positive R_{ijkl}. A sphere of radius r has an $R_{ijkl} = 1/r^2$ everywhere. The analytical meaning of the compatibility equations is the condition of integrability of the entire set of equations. For instance, if there is gravity, the Gaussian curvature is non-zero.

$$R_{ijkl} = 0 \tag{3.9}$$

$$\partial^3_{xyy}\varphi - \partial^3_{xyy}\varphi = 0 \tag{3.10}$$

$$\partial^3_{yxx}\varphi - \partial^3_{yyx}\varphi = 0 \tag{3.11}$$

In the physical sciences, the Airy function is a special function named after the British astronomer George Biddell Airy (1801–1892). To solve for the stress field, we introduce Airy function φ such that (Eq. 3.8),

$$\sigma_{xx} = \partial^2_{yy}\varphi(x,y) \quad \varepsilon_{xx} = \frac{1}{E}\left(\partial^2_{yy} - \vartheta\partial^2_{xx}\right)\varphi(x,y) \tag{3.12}$$

$$\sigma_{yy} = \partial^2_{xx}\varphi(x,y) \quad \varepsilon_{yy} = \frac{1}{E}\left(\partial^2_{xx} - \vartheta\partial^2_{yy}\right)\varphi(x,y) \tag{3.13}$$

$$\sigma_{xy} = \partial^2_{xy}\varphi(x,y) \quad \varepsilon_{xy} = -\frac{1+\vartheta}{E}\partial^2_{xy}\varphi(x,y) \tag{3.14}$$

This Airy function is constructed in such a way that the equilibrium equation is automatically satisfied. Now substituting into compatibility Eq. (3.7), we arrive at

$$\left[\partial^4_{xxxx} - \vartheta\partial^4_{xxyy} + \partial^4_{yyyy} - \vartheta\partial^4_{yyxx} + 2(1+\vartheta)\partial^4_{xyxy}\right]\frac{1}{E}\varphi(x,y) = 0 \tag{3.15}$$

By correcting Eq. (3.15) for Euclidean space,

$$\partial^2_{xy} = \partial^2_{yx} \tag{3.16}$$

the Laplacian operator in Eq. (3.15) in Cartesian coordinate is shown next.

$$\left[\partial^4_{xxxx} + 2\partial^4_{xxyy} + \partial^4_{yyyy} \right] = (\partial^2_{xx} + \partial^2_{yy})^2 \tag{3.17}$$

Using Laplacian operator Δ for right hand side of Eq. (3.15), we end up with the following biharmonic operator on function φ, finally leading to a fourth order partial differential equation. Theory of Elasticity is the only field of science that deals with a second-order Laplacian operator (a fourth-order partial differential equation).

$$\Delta^2 \varphi(x,y) = 0 \tag{3.18}$$

Later, more advanced techniques have been developed using complex variables. Employing complex variable methods enables many problems to be solved that would be intractable by other schemes. The method is based on the reduction of the elasticity boundary-value problem to a formulation in the complex domain. Complex variable theory provides a very powerful tool for the solution of many problems in elasticity. Such applications include solutions to the torsion problem and most importantly the plane problem. The technique is also useful for cases involving anisotropic and thermoelastic materials. This formulation then allows many powerful mathematical techniques available from complex variable theory to be applied to the elasticity problem.

But, today, most of these problems are addressed numerically. However, there are problems that cannot be resolved just numerically due to changes in singularity. Analytical solutions to elasticity problems are normally accomplished for regions and loadings with relatively simple geometry. However, most real-world problems involve structures with complicated shape and loading. This has led to the development of many numerical solution methods for elastic stress analysis. Over the past several decades, two methods have emerged that provide necessary accuracy, general applicability, and ease of use, and this has led to their wide acceptance by the stress analysis community. The first of these techniques is known as the finite element method (FEM) and involves dividing the body under study into several pieces or subdomains called elements. Because element size, shape, and approximating scheme can be varied to suit the problem, the method can accurately simulate solutions to problems of complex geometry and loading, and thus it has become a primary tool for practical stress analysis. The second numerical scheme, called the boundary element method (BEM), is based on an integral statement of elasticity. This statement may be cast into a form with unknowns only over the boundary of the domain under study. The boundary integral equation can then be solved using finite element concepts, and thus the method can accurately solve a large variety of problems. This chapter provides an overview of each method, focusing on narrow applications for two-dimensional elasticity problems. The primary goal is to establish a basic level of understanding that will allow a quick look at applications and enable connections to be made between numerical solutions (simulations) and those developed analytically in the previous chapters.

Few principles are also important to address complex geometry elasticity problems. These are (a) superposition principle and (b) Saint Venant's principle. Superposition principle allows breaking down a complex stress state into simple solvable pieces and linearly combining. The principle of superposition simply states that on a linear elastic structure, the combined effect of several loads acting simultaneously is equal to the algebraic sum of the effects of each load acting individually. For example, this principle implies, for a beam, that the total reactions due to the two loads acting simultaneously could have been obtained by algebraically summing, or superimposing, the reactions due to each of the two loads acting individually. The principle of superposition considerably simplifies the analysis of structures subjected to different types of loads acting simultaneously and is used extensively in structural analysis. The principle is valid for structures that satisfy the following two conditions: (a) The deformations of the structure must be so small that the equations of equilibrium can be based on the undeformed geometry of the structure; and (b) the structure must be composed of linearly elastic material; that is, the stress–strain relationship for the structural material must follow Hooke's law. The structures that satisfy these two conditions respond linearly to applied loads and are referred to as linear elastic structures. Engineering structures are generally designed so that under service loads, they undergo small deformations with stresses within the initial linear portions of the stress–strain curves of their materials. Thus, most common types of structures under service loads can be classified as linear elastic; therefore, the principle of superposition can be used in their analysis.

Saint Venant's principle, named after Adhémar Jean Claude Barre de Saint-Venant, states that the elastic fields (stress, strain, displacement) resulting from two different, but statically equivalent loading conditions, are approximately the same everywhere except in the vicinity of the point of application of the load. The original statement was published in French in 1855. The Saint-Venant's principle allows one to replace complicated stress distributions or weak boundary conditions with ones that are easier to solve, as long as that boundary is geometrically short. According to this principle, high-order moment of mechanical load (moment with order higher than torque) decays so fast that it never needs to be considered for regions far from the short boundary. Therefore, the principle of Saint-Venant can be regarded as a statement on the asymptotic behavior of the Green's function by a point-load. Green's function is the impulse response of an inhomogeneous linear differential operator defined on a domain with specified initial conditions or boundary conditions. This means that if $\mathcal{L}$ is the linear differential operator, then the Green's function G is the solution of the equation $\mathcal{L}G = \delta$, where δ is Dirac's delta function.

Chapter 4

Conventional Fracture Mechanics

Fracture mechanics is the discipline concerned with analyzing the failure of materials containing cracks and flaws. With Fracture mechanics, we can determine the stress level at which cracks of known size can propagate through the material to final failure. Now that we have fundamental equation for linear elastic stress field in previous chapter, we need to introduce a crack in such a medium. Figure 4.1 shows single-edge notch crack in an infinite elastic plane with elements showing stress tensor in X–Y plane and angularly rotated r, θ plane. Fracture mechanics was developed during World War I by an English aeronautical engineer A.A. Griffith—thus the term Griffith crack—to explain the failure of brittle materials. Griffith's work was motivated by two contradictory facts: (a) The breaking strength is much lower than stress needed for breaking atomic bonds of glass and (b) experiments on glass fibers that Griffith himself conducted suggested that the fracture stress increases as the fiber diameter decreases. Hence, the uniaxial tensile strength, which had been used extensively to predict material failure before Griffith, could not be a specimen-independent material property. Griffith suggested that the low fracture strength observed in experiments, as well as the size-dependence of strength, was due to the presence of microscopic flaws in the bulk material. To verify the flaw hypothesis, Griffith introduced an artificial flaw in his experimental glass specimens. The artificial flaw was in the form of a surface crack which was much larger than other flaws in a specimen. The experiments showed that the product of the square root of the flaw length (ℓ) and the stress at fracture was nearly constant.

An explanation of this relation in terms of linear elasticity theory is problematic. Linear elasticity theory predicts that stress (and hence the strain) at the tip of a sharp flaw in a linear elastic material is infinite. To avoid that problem, Griffith developed a thermodynamic approach to explain the relation that he observed. The growth of a crack, the extension of the surfaces on either side of the crack, requires an increase in the surface energy. Griffith found an expression for the constant λ in terms of the surface energy of the crack by solving the elasticity problem of a finite crack in an elastic plate. Briefly, the approach was to compute the potential energy stored in a perfect

DOI: 10.1201/9781003359845-5

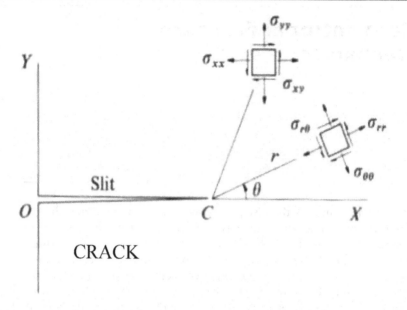

Figure 4.1 Stress distribution around a crack displayed by the rectangular elements in X–Y and polar coordinate.

specimen under a uniaxial tensile load. Then fix the boundary so that the applied load does no work and then introduce a crack into the specimen. The crack relaxes the stress and hence reduces the elastic energy near the crack faces. On the other hand, the crack increases the total surface energy of the specimen. Finally, the change in the free energy is computed (surface energy – elastic energy) as a function of the crack length. Failure occurs when the free energy attains a peak value at a critical crack length, beyond which the free energy decreases as the crack length increases, i.e., by causing fracture. Using this procedure, in 1920s, A.A. Griffith proposed an energy-based crack growth criterion that states that for the crack to grow, two surfaces need to be created. If the surface energy is γ, then critical energy release rate G_c for a crack to grow can be simply expressed as $G_c = 2\gamma$ (creation of two surfaces). Griffith validated his theory through experimental work using inorganic glass and formed the foundation for the role of defects on strength of materials [Griffith, A.A. (1921), Broek, D. (1974), Parker, A.P. (1981)]. Griffith was able to experimentally measure surface energy by heating glass up to a semi-liquid state. G in the aforementioned equation is strain energy release rate for an incremental crack extension and can be related to applied stress σ_∞ as $G = \frac{\pi}{E} \sigma_\infty{}^2 l$ for plane stress condition and $G = \frac{\pi}{E} \sigma_\infty{}^2 l (1 - \vartheta^2)$ for plane strain conditions. In continuum mechanics, a material is said to be under plane stress if the stress vector is zero across a particular plane. When that situation occurs over an entire element of a structure, as is often the case for thin plates, the stress analysis is considerably simplified, as the stress state can be represented by a tensor of dimension 2 (representable as a 2×2 matrix rather than 3×3). Plane stress typically occurs in thin flat plates that are acted upon only by load forces that are parallel to them. In certain situations, a gently thin curved plate may also be assumed to have plane stress for the purpose of stress analysis. This is the case, for example, of a thin-walled cylinder filled with a fluid under pressure. In such cases, stress components perpendicular to the plate are negligible compared to those parallel to it. In other situations, however, the bending stress of a thin plate cannot be neglected. One can still simplify the analysis by using a two-dimensional domain, but the plane stress tensor at each point must be complemented with bending terms. Plane strain is the physical deformation of a body that occurs when the material is displaced in a direction parallel to a plane. This situation applies to thick plates. G here is not the same as Gibb's free energy discussed in later sections. Also, energy release rate is a vector while Gibb's energy is a scalar.

Subsequently, efforts began in solving stress field around a crack to extend Griffith's fracture criterion to engineering systems under complex stress field. To solve the stress field in front of the crack tip, it is preferable to rotate in the angular plane (r, θ), i.e.,

$$\varphi(x, y) \rightarrow \varphi(r, \theta) \tag{4.1}$$

By definition, all the stress components in r, θ space can be written as

$$\sigma_{rr} = \left(\frac{1}{r} \partial_r + \frac{1}{r^2} \partial_\theta^2 \right) \varphi(r, \theta) \tag{4.2}$$

$$\sigma_{\theta\theta} = \partial_{rr}^2 \varphi(r, \theta) \tag{4.3}$$

$$\sigma_{r\theta} = -\partial_r \left(\frac{1}{r} \partial_\theta \varphi(r, \theta) \right) \tag{4.4}$$

Now we need to use biharmonic operator on $\varphi(r, \theta)$ since proper formulation of the physical problem should be formed in invariant terms.

$$\Delta^2 \varphi(r, \theta) = 0 \tag{4.5}$$

Δ in polar coordinate is shown here.

$$\Delta = \left\{ \partial_{rr}^2 + \frac{1}{r} \partial_r + \frac{1}{r^2} \partial_{\theta\theta}^2 \right\} \tag{4.6}$$

Before introducing crack in an elastic medium, we need to introduce the concept of self-similarity (*not mentioned in mechanics books*). Imagining a crack in an elastic medium with stress applied at infinity, we need to introduce a distance r from the crack tip at any angular position θ with respect to the crack plane. Self-similarity requires that if one changes the distance scale r, the relationship between $\varphi(r,\theta)$ and $\varphi(c_1 r,\theta)$ holds as $c_2(c_1)\varphi(c_1 r,\theta) = \varphi(r,\theta)$. Consequently,

$$\frac{dc_2(c_1)}{dc_1} * \varphi(c_1 r,\theta) + c_2(c_1) * \frac{\partial \varphi(c_1 r,\theta)}{\partial(c_1 r)} * r = 0 \tag{4.7}$$

$$\frac{dc_2(c_1)}{dc_1} * \varphi(c_1 r,\theta) + \frac{c2(c_1)}{c_1} * \frac{\partial \varphi(c_1 r,\theta)}{\partial(c_1 r)} * c_1 r = 0 \tag{4.8}$$

$$\frac{c_1 dc_2(c_1)}{c_2 dc_1} = -\frac{\partial \varphi(c_1 r,\theta) c_1 r}{\partial(c_1 r)\varphi(c_1 r,\theta)} \tag{4.9}$$

$$\frac{\partial \ln\varphi(r,\theta)}{\partial \ln r} = -\lambda \tag{4.10}$$

Right hand side of the equation is independent of θ and represented by λ. Solution of Eq. (4.8) can be presented as

$$\ln\varphi(r,\theta) = -\lambda \ln r + \ln\phi(\theta) \tag{4.11}$$

Please note that $\ln\phi(\theta)$ is independent of r. We can then reduce Eq. (4.10) to

$$\ln\varphi(r,\theta) = \ln\left(r^{-\lambda}\phi(\theta)\right) \tag{4.12}$$

$$\varphi(r,\theta) = r^{-\lambda}\phi(\theta) \tag{4.13}$$

Thus, the dependence on r and θ is split on taking the inverse operation. This function satisfies self-similarity and is a very useful representation and can be treated with the Laplacian operator in Eq. 4.6 discussed earlier.

$$\partial_{rr}^2 \varphi(r,\theta) = \lambda(\lambda+1)r^{-(\lambda+2)}\phi(\theta) \tag{4.14}$$

$$\frac{1}{r}\partial_r \varphi(r,\theta) = -\lambda r^{-(\lambda+2)}\phi(\theta) \tag{4.15}$$

$$\frac{1}{r^2}\partial_{\theta\theta}^2 \varphi(r,\theta) = r^{-(\lambda+2)}\phi''(\theta) \tag{4.16}$$

Considering that Δ^2 of $\varphi(r,\theta)$ equals zero from the compatibility Eq. (4.6), Δ of the right-hand side of the equation should reduce to zero.

$$\Delta\left[\lambda(\lambda+1)\phi(\theta) - \lambda\phi(\theta) + \phi''(\theta)\right]r^{-(\lambda+2)} = 0 \tag{4.17}$$

Second derivative on $r^{-(\lambda+2)}$ yields following equation.

$$\partial^2_{rr} r^{-(\lambda+2)} = (\lambda+2)(\lambda+3) r^{-(\lambda+2)} \tag{4.18}$$

Continuing with the operation on Eq. (4.17), following expression emerges.

$$\lambda^2 (\lambda+2)^2 \phi(\theta) + \left[\lambda^2 + (\lambda+2)^2 \right] \phi''(\theta) + \phi^{IV}(\theta) = 0 \tag{4.19}$$

Next few steps are related to determining λ from boundary value problem associated with a sharp crack in an elastic medium. As shown in Figure 4.2, stress approaches infinity at the crack tip—a typical singularity problem in Physics. Following Euler problem, we can define $\phi(\theta)$ and its derivatives as given in Eq. (4.20).

$$\phi(\theta) = \epsilon^{k\theta} \tag{4.20}$$

$$\phi''(\theta) = k^2 \epsilon^{k\theta} \tag{4.21}$$

$$\phi^{IV}(\theta) = k^4 \epsilon^{k\theta} \tag{4.22}$$

Substituting Eqs. (4.20–4.22) into (4.19), we end up with an algebraic equation in k and λ.

$$\lambda^2 (\lambda+2)^2 + \left[\lambda^2 + (\lambda+2)^2 \right] k^2 + k^4 = 0 \tag{4.23}$$

Replacing k^2 with m, we can further simplify 56 as

$$m^2 + \left[\lambda^2 + (\lambda+2)^2 \right] m + \lambda^2 (\lambda+2)^2 = 0 \tag{4.24}$$
for $k = \pm\sqrt{m}$

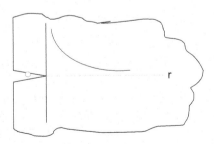

Figure 4.2 Stress approaches infinity at the crack tip.

Equation (4.22) is a quadratic equation of the form $X^2 + pX + q = 0$. One can utilize Vieta's theorem from elementary algebra to find the roots X_1 and X_2.

$$p = -(X_1 + X_2)$$
$$q = X_1 X_2$$

Thus, from Eq. 4.24, the roots m_1 and m_2 can be derived as

$$m_1 = -\lambda^2$$
$$m_2 = -(\lambda + 2)^2$$

Since $k = \pm\sqrt{m}$, one can express components of k as

$$k_{1,2} = \pm\sqrt{-\lambda^2} = \pm\lambda i \tag{4.25}$$
$$k_{3,4} = \pm\sqrt{-(\lambda + 2)^2} = \pm(\lambda + 2)i \tag{4.26}$$

Application of k to Eqs. (4.20–4.22) results in four linearly independent Euler equations.

$$\phi_1(\theta) = e^{\lambda\theta i}$$
$$\phi_2(\theta) = e^{-\lambda\theta i}$$
$$\phi_3(\theta) = e^{(\lambda+2)\theta i}$$
$$\phi_4(\theta) = e^{-(\lambda+2)\theta i} \tag{4.27}$$

A general solution can be written as a sum of the aforementioned solutions in 60.

$$\phi(\theta) = c_1 e^{\lambda\theta i} + c_2 e^{-\lambda\theta i} + c_3 e^{(\lambda+2)\theta i} + c_4 e^{-(\lambda+2)\theta i} \tag{4.28}$$

Knowing that $e^{i(\pm\lambda\theta)} = \cos\lambda\theta \pm i\sin\lambda\theta$, one can rewrite Eq. (4.27) in terms of angular dimension as

$$\phi(\theta) = A\cos\lambda\theta + B\sin\lambda\theta + C\cos(\lambda + 2)\theta + D\sin(\lambda + 2)\theta \tag{4.29}$$

where A, B, C, and D are constants that incorporate $\pm i$. Since cos is a symmetric function and sin is antisymmetric. At this point, we need to introduce various modes of fracture since only portions of the aforementioned solution are applicable to each mode. Figure 4.3 schematically represents three modes of fracture. Mode I represents the crack opening mode, mode II represents the crack shearing mode, and finally mode III represents crack tearing mode.

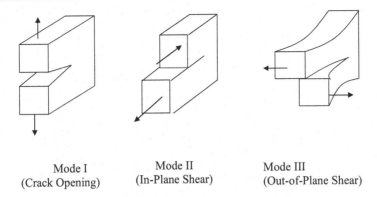

Mode I   Mode II   Mode III

(Crack Opening)   (In-Plane Shear)   (Out-of-Plane Shear)

Figure 4.3 Schematics of three primary modes of fracture.

Equation (4.29) can thereby split into two equations where Eq. (4.30) represents mode I and Eq. (4.31) represents mode II.

$$\phi_+(\theta) = A\cos\lambda\theta + C\cos(\lambda+2)\theta \tag{4.30}$$

$$\phi_-(\theta) = B\sin\lambda\theta + D\sin(\lambda+2)\theta \tag{4.31}$$

Figure 4.4 is a planer view of a body under external tensile forces F presented defining the angle θ with respect to the axis of symmetry (Eq. 4.30). Similarly, Figure 4.5 is a planer view of a body under external shear forces F presented defining the angle θ with respect to the axis of symmetry (Eq. 4.31).

Mode III is an out-of-plane problem, and none of these equations are applicable. Going back to Eqs. (4.1–4.3), one can deduce that $\varphi_+ = r^{-\lambda}\phi_+$. Comparing Eqs. (4.2–4.4) and (4.14–4.16), one can deduce the following set of equations.

$$\sigma_{\theta\theta} = \lambda(\lambda+1)r^{-(\lambda+2)}\phi_+(\theta) \tag{4.32}$$

$$\sigma_{r\theta} = -\partial_r\left\{\frac{1}{r}r^{-\lambda}\partial_\theta\phi_+(\theta)\right\} \tag{4.33}$$

Finally, Eq. (4.33) reduces to

$$\sigma_{r\theta} = (\lambda+1)r^{-(\lambda+2)}\phi_+(\theta) \tag{4.34}$$

Symmetry requires either $\sigma_{\theta\theta}(r,\theta)\big|_{\theta=\pm\infty} = 0$ or $\sigma_{\theta\theta}(r,\theta)\big|_{\theta=\infty} = 0$. This implies in Eq. (4.33), θ can be replaced by ∞. Applying this to Eqs. (4.30) and (4.31), we arrive at

$$A\cos\lambda\infty + C\cos(\lambda+2)\infty = 0 \tag{4.35}$$

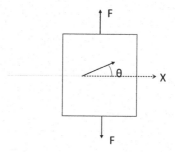

Figure 4.4 Force F is perpendicular to the axis of symmetry in mode I.

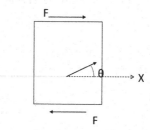

Figure 4.5 Force F is parallel to the axis of symmetry in mode II.

$$-A\lambda sin\lambda\infty - C(\lambda+2)sin(\lambda+2)\infty = 0 \qquad (4.36)$$

To find a nontrivial solution to Eqs. (4.35) and (4.36), determinant should reduce to zero. Thus,

$$\begin{vmatrix} cos\lambda\infty & cos(\lambda+2)\infty \\ \lambda sin\lambda\infty & (\lambda+2)sin(\lambda+2)\infty \end{vmatrix} = 0 \qquad (4.37)$$

Left hand side of Eq. (4.37) reduces to

$$(\lambda+2)cos\lambda\infty sin(\lambda+2)\infty - \lambda cos(\lambda+2)\infty sin\lambda\infty = 0 \qquad (4.38)$$

$$\lambda[cos\lambda\infty sin(\lambda+2)\infty - cos(\lambda+2)\infty sin\lambda\infty] + 2cos\lambda\infty sin(\lambda+2)\infty = 0 \quad (4.39)$$

One can replace the term inside "[]" with $sin(\lambda+2-\lambda)\infty$.

$$[\lambda sin2\infty + 2sin\lambda\infty cos\lambda\infty cos2\infty] + 2cos^2\lambda\infty sin2\infty = 0 \qquad (4.40)$$

Resolving further,

$$sin2\infty(\lambda + 2cos^2\lambda\infty) + sin2\lambda\infty cos2\infty = 0 \qquad (4.41)$$

If $cos2\infty \neq 0$

$$tan2\infty = \frac{-sin2\lambda\infty}{\lambda + 2cos^2\lambda\infty} \tag{4.42}$$

Considering the axis of symmetry in Figures 4.4 and 4.5, infinity ∞ reduces to π.

Equation (4.42) further reduces to

$$\frac{sin2\lambda\pi}{\lambda + 2cos^2\lambda\pi} = 0 \quad tan2\pi = 0 \tag{4.43}$$

Since denominator definitely cannot be zero, we deduce

$$sin2\lambda\pi = 0 \tag{4.44}$$

$$2\lambda\pi = n\pi, n = 0, \pm1, \pm2, \pm3, \pm4, \ldots\ldots.. \tag{4.45}$$

$$\lambda = \frac{n}{2}, \lambda = 0, \pm\frac{1}{2}, \pm1, \pm\frac{3}{2}, \pm2, \ldots\ldots \tag{4.46}$$

Reverting back to the expression of stress components where λ appears,

$$\varphi(r,\theta) = r^{-(\lambda+2)}\phi_+(\theta) \tag{4.47}$$

We could express the system of Eqs. (4.27) in more expanded form as given here.

$$\begin{pmatrix} \sigma_{rr} \\ \sigma_{\theta\theta} \\ \sigma_{r\theta} \end{pmatrix} = r^{-(\lambda+2)} \begin{pmatrix} \phi_{rr}(\theta) \\ \phi_{\theta\theta}(\theta) \\ \phi_{r\theta}(\theta) \end{pmatrix} \tag{4.48}$$

In order to determine λ, let us consider $(\lambda+2)$ as > 0, $= 0$, and < 0.

For $(\lambda+2) \leq 0$,

If $(\lambda+2) = 0$, stresses lose any physical meaning.

If $(\lambda+2) < 0$ leads to positive $-(\lambda+2)$, stresses will approach ∞ far away from crack which does not hold any physical meaning. This leaves us with $(\lambda+2) > 0$. Finally, we arrive at following values for λ.

$$\lambda = -\frac{3}{2}, -1, -\frac{1}{2}, 0, \frac{1}{2}, 1, \frac{3}{2}, 2, \ldots\ldots\ldots \tag{4.49}$$

One notices that only first solution has $-\frac{3}{2}$ real physical meaning. This leads to the solution for elastic stress field ahead of the crack tip.

$$\begin{pmatrix} \sigma_{rr} \\ \sigma_{\theta\theta} \\ \sigma_{r\theta} \end{pmatrix} = r^{-\left(\frac{1}{2}\right)} \begin{pmatrix} \phi_{rr}(\theta) \\ \phi_{\theta\theta}(\theta) \\ \phi_{r\theta}(\theta) \end{pmatrix} \tag{4.50}$$

In the theory of elasticity, we do not consider the rest of the spectrum of numbers for λ due to the requirement of finite energy and finite displacement.

I. $\lim\limits_{r\to\infty} \Delta\tilde{\sigma}' = 0$

Based on the aforementioned requirement, all values of λ in 82 is acceptable. Next requirement is physical not mathematical. Consider $\varphi(r,\theta) = r^{-\lambda}\phi(\theta)$.

II. Limited Energy, $\tilde{\sigma} \sim r^{-(\lambda+2)}$ and $\tilde{\varepsilon} \sim r^{-(\lambda+2)}$

Elastic energy scales with distance from the crack tip as $\rho \sim \dfrac{1}{2}\tilde{\sigma}\tilde{\varepsilon} \sim r^{-2(\lambda+2)}$.

Now considering the A area around the crack tip, elastic energy density expressed here should be limited.

$$\int_A \rho(r,\theta)\,rd\theta dr = \int_{-\pi}^{\pi} d\theta \int_0^R \rho(r,\theta)\,rdr \tag{4.51}$$

Considering the second integral only, we can express ρ in terms of λ as

$$\int_0^R r^{-2(\lambda+2)}rdr = \int_0^R r^{-2(\lambda+2)+1}dr \tag{4.52}$$

Equation (4.33) further reduces on integration using R, a distance from crack tip.

$$\frac{1}{-2(\lambda+2)+2}r^{-2(\lambda+2)+2}\Big|_0^R \tag{4.53}$$

The aforementioned equation holds true only when $-2(\lambda+2) > 0$ leading to $\lambda < -1$. This helps establish again that only $\left(\lambda = -\dfrac{3}{2}\right)$ satisfies both requirements. Equation (4.28) can be rewritten as

$$\varphi(r,\theta) = r^{-\frac{3}{2}}\phi(\theta) \tag{4.54}$$

Recalling from earlier, $\phi(\theta)$ can be expressed as a cosine symmetry function

$$\phi(\theta) = Acos\left(-\frac{3}{2}\theta\right) + Bcos\left(-\frac{3}{2}+2\right)\theta = Acos\frac{3}{2}\theta + Bcos\frac{\theta}{2} \tag{4.55}$$

Stress components can be rewritten by introducing 4.44 into 4.37 where A and B are constants.

$$\sigma_{rr} = \frac{3}{2}r^{-\frac{1}{2}}\left(Acos\frac{3}{2}\theta + Bcos\frac{\theta}{2}\right) - r^{-\frac{1}{2}}\left(\frac{9}{4}Acos\frac{3}{2}\theta + \frac{1}{4}Bcos\frac{\theta}{2}\right)$$
$$= r^{-\frac{1}{2}}\left(-\frac{3}{4}Acos\frac{3}{2}\theta + \frac{5}{4}Bcos\frac{\theta}{2}\right) \tag{4.56}$$

$$\sigma_{\theta\theta} = \frac{3}{2}\frac{1}{2}r^{-\frac{1}{2}}\left(Acos\frac{3}{2}\theta + Bcos\frac{\theta}{2}\right) = \frac{3}{4}r^{-\frac{1}{2}}\left(Acos\frac{3}{2}\theta + Bcos\frac{\theta}{2}\right) \tag{4.57}$$

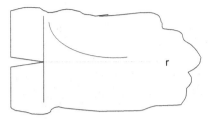

Figure 4.6 Crack of length ℓ subjected to stress σ_∞ at infinity in mode I.

$$\sigma_{r\theta} = \frac{1}{2}r^{-\frac{1}{2}}\left(\frac{3}{2}A\sin\frac{3}{2}\theta + \frac{1}{2}B\sin\frac{\theta}{2}\right)$$ (4.58)

Let us consider a crack in an infinite medium with stress singularity at the tip of the crack (Figure 4.6). Let us attempt to assess A and B using dimensional analysis. Dimension of stress is Nm^{-2}, and dimension of r is m. This leads to dimension of A and B to be $m^{\frac{1}{2}}[\sigma]$, $Nm^{-\frac{3}{2}}$, respectively. Again, considering stress scales with $[A,B].r^{-\frac{1}{2}}$, let us now consider crack of length ℓ in an infinite elastic medium under stress at infinity, σ_∞. Dimensional analysis leads to $K = \sigma_\infty\sqrt{\pi\ell}$. Finally, we reach the equation for stress components for a crack opening mode I expressed in polar coordinates.

$$\begin{pmatrix}\sigma_{rr}\\ \sigma_{\theta\theta}\\ \sigma_{r\theta}\end{pmatrix} = \frac{K_I}{\sqrt{2\pi r}}\begin{pmatrix}a_1\cos\frac{3}{2}\theta + b_1\cos\frac{\theta}{2}\\ a_2\cos\frac{3}{2}\theta + b_2\cos\frac{\theta}{2}\\ a_3\sin\frac{3}{2}\theta + b_3\sin\frac{\theta}{2}\end{pmatrix}$$ (4.59)

Similar derivations are possible with modes II and III yielding the following pair of equations: $\begin{pmatrix}\sigma_{rr}\\ \sigma_{\theta\theta}\\ \sigma_{r\theta}\end{pmatrix} = \frac{K_{II}}{\sqrt{2\pi r}}(f(\theta))$ and $\begin{pmatrix}\sigma_{rr}\\ \sigma_{\theta\theta}\\ \sigma_{r\theta}\end{pmatrix} = \frac{K_{III}}{\sqrt{2\pi r}}(f(\theta))$

Based on the aforementioned equations one can argue that, if one accepts the linear Fracture mechanics postulates of small strain and linear stress–strain relationship, slow crack growth is impossible due to crack instability. The application of an infinitely sharp crack will lead to instantaneous failure ("enigma of linear elastic Fracture mechanics"). Since we know from experimental observations that slow crack growth does take place, the solution of that enigma is in the Crack layer theory where the damage accompanying the crack resolves the paradox. Conventional Fracture mechanics deal with this situation in empirical ways. In a few more sections, we will delve into Crack layer theory that will incorporate irreversible thermodynamics to establish kinetic equations for slow crack growth.

Fracture Mechanics, in its most general interpretation, comprises all modes of failure, including buckling, large deformation and rupture (ductile fracture), failure due to a distributed damage growth, as well as a brittle fracture. Ideally brittle fracture occurs by crack propagation with elastic (reversible) deformation of the surrounding material. Thus, the two surfaces of the crack perfectly match when put together. The fracture of common window glass illustrates the ideally brittle fracture. Brittle fracture in most of engineering materials is different from the ideally brittle one: There is a localized irreversible deformation along crack faces and the fracture surfaces do not match when put together, but crack growth is still the dominant feature of fracture process. In this book, we use the term Fracture mechanics in a narrow sense, as the mechanics of cracks. For about a century, Fracture mechanics has been evolving in two main directions. One of them deals with the determination of the stress, strain, and displacement fields for cracks in elastic, elastoplastic, and visco-elastoplastic materials. The pioneering work of Inglis (1913) as mentioned earlier marks the beginning of computational Fracture mechanics. Analytical and numerical modeling of such problems has become a mature science that is enhanced by recent advances in computational capabilities. Another direction of Fracture mechanics is associated with an understanding of physics of fracture which has not been addressed adequately other than through pioneering work of current authors. In particular, the following aspects should be mentioned: Mechanism(s) and criteria of crack initiation; condition(s) of crack equilibrium; observations of slow crack growth (SCG) under different loading conditions (creep, fatigue, impact, etc.); formulation of the kinetic equations of SCG; observations of rapid (dynamic) crack propagation and formulation of the dynamic crack growth equations; and criteria of transition from slow-to-rapid growth. One should also mention statistical aspects of brittle fracture manifested in a large scatter of fracture parameters, stochastic character of crack trajectories, various scale effects, etc. Some of these issues will be discussed in subsequent chapters.

Ultimately, Fracture mechanics is expected to provide quantitative answers to the following questions:

- What is the strength of the component as a function of crack size?
- What crack size can be tolerated under service loading, i.e., what is the maximum permissible crack size?
- How long does it take for a crack to grow from a certain initial size, for example the minimum detectable crack size, to the maximum permissible crack size?
- What is the service life of a structure when a certain preexisting flaw size (e.g., a manufacturing defect) is assumed to exist?
- During the period available for crack detection, how often should the structure be inspected for cracks?

The beginning of physics of fracture studies can be attributed to Griffith (1921) who pointed out that all engineering materials contain a population of defects, like small cracks, that cause a significant reduction of actual strength, as compared to the theoretical strength. Griffith also suggested that, in order for a crack to grow, a release of the elastic strain energy associated with crack extension should be equal to (or larger than) the energy needed for the formation of new crack surface and formulated the crack equilibrium condition (Griffith, 1921, 1925). Since Griffith's time, the majority of studies have been dedicated to the observation and characterization of SCG in various engineering materials and formulation of empirical crack growth equation. Much less work has been done dealing with rapid crack propagation, and even fewer number of publications address the mechanisms and criteria of crack initiation. Thus, in contrast with Computational Fracture mechanics, the physics of brittle fracture is still far from the mature state. One can distinguish three main stages of brittle fracture: (a) initiation of a small crack as a precursor of a macroscopic crack, (b) slow crack growth, and (c) instability and transition to a rapid (dynamic) crack propagation (RCP). The duration of stage (c) is very short (typically milliseconds) since rapid crack in most engineering materials propagates with the speed of a few hundred meters per second or faster. Therefore, the lifetime t_f of a structural element comprises the duration of the crack-initiation stage t_n and the stable slow crack growth t_p. Griffith's work was largely ignored by the engineering community until the early 1950s. The reasons for this appear to be that (a) in the actual structural materials, the level of energy needed to cause fracture is orders of magnitude higher than the corresponding surface energy and (b) in structural materials, there are always some inelastic deformations around the crack front, which would make the assumption of linear elastic medium with infinite stresses at the crack tip highly unrealistic.

Twenty-five to thirty years later from Griffith's first finding and after numerous catastrophic failures, the basic elements of contemporary Fracture mechanics were developed in 1957 by George R. Irwin, the man usually considered to be the father of Fracture mechanics [Irwin, G.R. (1957)]. He introduced the stress intensity factor (SIF) represented by the variable, K. K_I introduced in Eq. (4.59) represents the SIF for mode I crack opening. It is one of the most fundamental and useful parameters in all Fracture mechanics. SIF depends on crack size, specimen shape and size, and applied load. Thus, a new field of science was born in mid-1900s addressing stress analysis of cracks called linear elastic Fracture mechanics (LEFM). Fracture mechanics uses methods of analytical solid mechanics to calculate the driving force on a crack and those of experimental solid mechanics to characterize the material's resistance to fracture. Theoretically, the stress ahead of a sharp crack tip becomes infinite and cannot be used to describe the state around a crack. Fracture mechanics is used to characterize the loads on a crack, typically using a single parameter to describe the complete loading state at the crack tip. A number of different parameters have been developed. When the plastic zone at the tip of the crack is small relative to the crack length, the stress state at the crack tip is the result of elastic forces within the material and is termed linear elastic Fracture mechanics (LEFM) and can be characterized using the stress intensity factor K. Although the load on a crack can be arbitrary, in 1957, G. Irwin found any state could be reduced to a combination of three independent stress intensity factors:

Mode I—Opening mode (a tensile stress normal to the plane of the crack),
Mode II—Sliding mode (a shear stress acting parallel to the plane of the crack and perpendicular to the crack front), and
Mode III—Tearing mode (a shear stress acting parallel to the plane of the crack and parallel to the crack front).

It is a theoretical construct usually applied to a homogeneous, linear elastic material and is useful for providing a failure criterion for brittle materials [Broek, D.; Parker, A.P.]. K is related to strain energy release rate using the relationships discussed earlier as $G = \pi/_E \sigma_\infty^2 \ell = \dfrac{K^2}{E}$ for plane stress and $G = \dfrac{K^2}{E}(1 - v^2)$ for plane strain conditions.

Another significant achievement of Irwin and his colleagues was to find a method of calculating the amount of energy available for fracture in terms of the asymptotic stress and displacement fields around a crack front in a linear elastic solid. This asymptotic expression for the stress field in mode I loading is related to the stress intensity factor. Irwin called the quantity the stress intensity factor. Since the quantity is dimensionless, the stress intensity factor can be expressed in units of MPa sqrt. Stress intensity replaced strain energy release rate, and a term called fracture toughness replaced surface weakness energy. Both terms are simply related to the energy terms that Griffith used. When $K = K_c$, fracture occurs. For the special case of plane strain deformation, K_c is considered a material property. K_{Ic}, K_{IIc}, and K_{IIIc} arise because of the different ways of loading a material to enable a crack to propagate. It refers to so-called "mode I, mode II and mode III".

The expression for K_I will be different for geometries other than the center-cracked infinite plate. Thus, it is necessary to introduce a dimensionless correction factor, Y, in order to characterize the geometry. This correction factor, also often referred to as the geometric shape factor, is given by empirically determined series and accounts for the type and geometry of the crack or notch. We thus have $K_I = Y.\sigma \sqrt{\pi \ell}$ where Y is a function of the crack length and width of sheet given, for a sheet of finite width. In the case of a sheet of finite width W containing a through-thickness crack of length $2a$, Y is given by $Y = \sqrt{sec(\pi a / W)}$. By solving boundary value problems for many other possible geometries, Tada et al. (1985) published results for Y.

Transition flaw size is important for failure stress as a function of crack size. Let a material have a yield strength σ_y and a fracture toughness in mode I, K_{Ic}. Based on Fracture mechanics, the material will fail at stress $\sigma_{fail} = K_{Ic} / \sqrt{\pi \ell}$. Based on plasticity, the material will yield when $\sigma_{fail} = \sigma_y$. These curves intersect when $\ell = K_{Ic}^2 / \sigma_y^2 \pi$. This value of ℓ is called as transition flaw size; $\ell(t)$ depends on the material properties of the structure. When $\ell < \ell(t)$, the failure is governed by plastic yielding, and when $\ell > \ell(t)$, the failure is governed by Fracture mechanics. The value $\ell(t)$ for engineering alloys is 100 mm, and for ceramics, it is 0.001 mm. If we assume that manufacturing processes can give rise to flaws in the order of micrometers, then it can be seen that ceramics are more likely to fail by fracture, whereas engineering alloys would fail by plastic deformation.

The characterizing parameter describes the state of the crack tip which can then be related to experimental conditions to ensure similitude. Crack growth occurs when the parameters typically exceed certain critical values. Corrosion may cause a crack to slowly grow when the stress corrosion–stress intensity threshold is exceeded. Similarly, small flaws may result in crack growth when subjected to cyclic loading known as fatigue; as it was claimed that for long cracks, the rate of growth is largely governed by the range of the stress intensity ΔK experienced by the crack due to the applied loading. Fast fracture will occur when the stress intensity exceeds the fracture toughness of the material. The prediction of crack growth is at the heart of the damage tolerance mechanical design discipline. The processes of material manufacture, processing, machining, and forming may introduce flaws in a finished mechanical component. Arising from the manufacturing process, interior and surface flaws are found in all metal structures. Not all such flaws are unstable under service conditions. Fracture mechanics is the analysis of flaws to discover those that are safe (i.e., do not grow) and those that are liable to propagate as cracks and so cause failure of the flawed structure. Despite these inherent flaws, it is possible to achieve through damage tolerance analysis the safe operation of a structure. For materials highly deformed before crack propagation, the linear elastic Fracture mechanics formulation is no longer applicable, and an adapted model is necessary to describe the stress and displacement field close to crack tip, such as on fracture of soft materials. In theory, the stress at the crack tip where the radius is nearly zero would tend to infinity. This would be considered a stress singularity, which is not possible in real-world applications. For this reason, in numerical studies in the field of Fracture mechanics, it is often appropriate to represent cracks as round tipped notches, with a geometry-dependent region of stress concentration replacing the crack-tip singularity. In actuality, the stress concentration at the tip of a crack within real materials has been found to have a finite value but larger than the nominal stress applied at infinity (σ_∞) to the specimen.

Around 1960, when the fundamentals of linear elastic Fracture mechanics were fairly well established, researchers turned their attention to crack-tip plasticity. Linear elastic Fracture mechanics (LEFM) ceases to be valid when significant plastic deformation precedes failure. During a relatively short time (1960–1961), several researchers developed analyses to correct for yielding at the crack tip, including Irwin, Dugdale, Barenblatt, and Wells [23]. The Irwin plastic zone correction was a relatively simple extension of LEFM, while Dugdale and Barenblatt each developed somewhat more elaborate models on the basis of a narrow strip of yielded material at the crack tip. When the size of the plastic zone at the crack tip is too large, elastic–plastic Fracture mechanics can be used with parameters such as the Rice's J-integral (explained later) or the crack tip opening displacement (CTOD).

Griffith's theory provides an excellent agreement with experimental data for brittle materials such as glass. For ductile materials such as steel, although the relation still holds, the surface energy (γ) predicted by Griffith's theory is usually unrealistically high. A group working under G.R. Irwin at the U.S. Naval Research Laboratory (NRL) during World War II realized that plasticity must play a significant role in the fracture of ductile materials. In ductile materials (and even in materials that appear to be brittle), a plastic zone develops at the tip of the crack. As the applied load increases, the plastic zone increases in size until the crack grows and the elastically strained material behind the crack tip unloads. The plastic loading and unloading cycle near the crack tip leads to the dissipation of energy as heat. Hence, a dissipative term must be added to the energy balance relation devised by Griffith for brittle materials. In physical terms, additional energy is needed for crack growth in ductile materials as compared to brittle materials.

Irwin's strategy was to partition the energy into two parts: Dissipative energy and the surface energy (and any other dissipative forces that may be at work). The dissipated energy provides the thermodynamic resistance to fracture. Then, the total energy is $2\lambda + 2\lambda_p$ where λ is the surface energy and λ_p is the plastic dissipation (and dissipation from other sources) per unit area of crack growth. Griffith's work was largely ignored by the engineering community until the early 1950s. The reasons for this appear to be (a) in the actual structural materials, the level of energy needed to cause fracture is orders of magnitude higher than the corresponding surface energy, and (b) in structural materials, there are always some inelastic deformations around the crack front that would make the assumption of linear elastic medium with infinite stresses at the crack tip highly unrealistic. For brittle materials such as glass, the surface energy term dominates and for polymers close to the glass transition temperature, we have intermediate values.

Irwin was the first to observe that if the size of the plastic zone around a crack is small compared to the size of the crack, the energy required to grow the crack will not be critically dependent on the state of stress (the plastic zone) at the crack tip. In other words, a purely elastic solution may be used to calculate the amount of energy available for fracture. The energy release rate for crack growth or strain energy release rate may then be calculated as the change in elastic strain energy per unit area of crack growth. Irwin showed that for a mode I crack (opening mode), the strain energy release rate and the stress intensity factor are related by $G_{Ic} = K_{IC}^2 / E$ where E is the Young's modulus. Irwin also showed that the strain energy release rate of a planar crack in a linear elastic body can be expressed in terms of the mode I, mode II (sliding mode), and mode III (tearing mode) stress intensity factors for the most general loading conditions.

Next, Irwin adopted the additional assumption that the size and shape of the energy dissipation zone remain approximately constant during brittle fracture. This assumption suggests that the energy needed to create a unit fracture surface is a constant that depends only on the material. This new material property was given the name fracture toughness and designated G_{Ic}. Today, it is the critical stress intensity factor K_{Ic}, found in the plane strain condition, which is accepted as the defining property in linear elastic Fracture mechanics. The same process as described before for a single event loading also applies to cyclic loading. If a crack is present in a specimen that undergoes cyclic loading, the specimen will plastically deform at the crack tip and delay the crack growth. In the event of an overload or excursion, this model changes slightly to accommodate the sudden increase in stress from that which the material previously experienced. At a sufficiently high load (overload), the crack grows out of the plastic zone that contained it and leaves behind the pocket of the original plastic deformation. Now, assuming that the overload stress is not sufficiently high as to completely fracture the specimen, the crack will undergo further plastic deformation around the new crack tip, enlarging the zone of residual plastic stresses. This process further toughens and prolongs the life of the material because the new plastic zone is larger than what it would be under the usual stress conditions. This allows the material to undergo more cycles of loading. This idea can be illustrated further by the graph of aluminum with a center crack undergoing overloading event.

NRL researchers concluded that naval materials, e.g., ship-plate steel, are not perfectly elastic but undergo significant plastic deformation at the tip of a crack. One basic assumption in Irwin's linear elastic Fracture mechanics is small scale yielding, the condition that the size of the plastic zone is small compared to the crack length. However, this assumption is quite restrictive for certain types of failure in structural steels though such steels can be prone to brittle fracture, which has led to a number of catastrophic failures. Linear–elastic Fracture mechanics is of limited practical use for structural steels, and fracture toughness testing can be expensive. Most engineering materials show some nonlinear elastic and inelastic behavior under operating conditions that involve large loads. In such materials, the assumptions of linear elastic Fracture mechanics may not hold, that is, the plastic zone at a crack tip may have a size of the same order of magnitude as the crack size. The size and shape of the plastic zone may change as the applied load is increased and also as the crack length increases. Therefore, a more general theory of crack growth is needed for elastic–plastic materials that can account for the local conditions for initial crack growth which include the nucleation, growth, and coalescence of voids (decohesion) at a crack tip.

Historically, the first parameter for the determination of fracture toughness in the elastoplastic region was the crack tip opening displacement (CTOD) or "opening at the apex of the crack" indicated. This parameter was determined by Wells during the studies of structural steels, which due to the high toughness could not be characterized with the linear elastic Fracture mechanics model. He noted that, before the fracture happened, the walls of the crack were leaving, and that the crack tip, after fracture, ranged from acute to rounded off due to plastic deformation. In addition, the rounding of the crack tip was more pronounced in steels with superior toughness. There are a number of alternative definitions of CTOD. In the two most common definitions, CTOD is the displacement at the original crack tip and the 90-degree intercept. The latter definition was suggested by Rice and is commonly used to infer CTOD in finite element models of such. Note that these two definitions are equivalent if the crack tip blunts in a semicircle. Most laboratory measurements of CTOD have been made on edge-cracked specimens loaded in three-point bending. Early experiments used a flat paddle-shaped gage that was inserted into the crack; as the crack opened, the paddle gage rotated, and an electronic signal was sent to an x–y plotter. This method was inaccurate, however, because it was difficult to reach the crack tip with the paddle gage. Today, the displacement V at the crack mouth is measured, and the CTOD is inferred by assuming the specimen halves are rigid and rotate about a hinge point (the crack tip).

Realizing the limitations of LEFM, some advancement has been made in later years to apply SIF to materials with crack tip plasticity as well. As expected, stress analysis of cracks led to fracture criteria based on critical SIF, K_c. Critical SIF can be directly related to Griffith's G_c establishing a fundamental relationship between stress field and energy. Stress analysis of cracks under various geometric configurations was solved by various researchers both numerically and analytically for design engineers taking Fracture mechanics into consideration for increasing lifetime [Tada et al., 1985]. Various modes of fracture initiation were introduced into the Fracture mechanics vocabulary known as mode I, mode II, and mode III fracture. Subsequently, three SIFs were introduced as K_I, K_{II}, and K_{III}.

The vocabulary was further expanded to include energy release rate (ERR) associated with crack advance, crack (tip) opening displacement (COD or CTOD), and ERR for small scale yielding, etc. To avoid complicacy, Γ represents the path over which the integral (J-integral) is taken avoiding the yielded zone for small-scale yielding problems. Once more following the established approach, fracture criteria are expressed as critical values G_c, COD_c or $CTOD_c$, and J_c. Elastic–plastic Fracture mechanics applies to materials that exhibit time-independent, nonlinear behavior (i.e., plastic deformation). Two elastic–plastic parameters are introduced here are CTOD and the J contour integral. Both parameters describe crack-tip conditions in elastic–plastic materials, and each has been used as a fracture criterion. Critical values of CTOD or J give nearly size-independent measures of fracture toughness, even for relatively large amounts of crack-tip plasticity. There are limits to the applicability of J and CTOD, but these limits are much less restrictive than the validity requirements of LEFM. Interesting to note while mechanics has advanced significantly with further understanding, failure/fracture criterion followed the old tradition of defining a critical value of a defined parameter.

Thus far, all discussions around Fracture mechanics have been limited to materials that behave in pure elastic manner. Nevertheless, there must be some sort of mechanism or property of the material that prevents such a crack from propagating spontaneously. The assumption is that the plastic deformation at the crack tip effectively blunts the crack tip. This deformation depends primarily on the applied stress in the applicable direction (in most cases, this is the y-direction of a regular Cartesian coordinate system), the crack length, and the geometry of the specimen. To estimate how this plastic deformation zone extended from the crack tip, Irwin equated the yield strength of the material to the far-field stresses of the y-direction along the crack (x direction) and solved for the effective radius. From this relationship, and assuming that the crack is loaded to the critical stress intensity factor, Irwin developed the following expression for the idealized radius of the zone of plastic deformation at the crack tip presented in the next section.

In practice, most materials show the so-called plastic deformation at the crack tip. Thus, modifications were introduced to linear elastic Fracture mechanics concepts through plasticity criteria introduced earlier in the Review of Classical Strength of Materials section. George Irwin was the first to introduce plastic zone model assuming that on the application of stress, crack tip singularity is instantly broken when material adjacent to the tip reaches material yield stress σ_y (simplest plasticity criterion). This possibly results in a plastic zone of length r_p. Irwin proposed a correction for SIF by adding a parameter $f(r_p)$ to crack length ℓ [Irwin, G.R. (1958)]. Finally, Irwin model yields:

$$r_p = \frac{1}{\pi}\left(\frac{K}{\sigma_y}\right)^2 \tag{4.60}$$

An alternative approach to that of Irwin was proposed by Dugdale (1960) and later in equivalent form independently by Barenblatt (1962). When a significant region around a crack tip has undergone plastic deformation, other approaches can be used to determine the possibility of further crack extension and the direction of crack growth and branching. A simple technique that is easily incorporated into numerical calculations is the cohesive zone model method which is based on concepts proposed independently by Baren-blatt and Dugdale in the early 1960s. The relationship between the Dugdale–Barenblatt models and Griffith's theory was first discussed by Willis in 1967.

Dugdale assumes that the crack extends right through the plastic zone; thus, the crack length is increased from ℓ to $\ell + \rho$, where ρ is extent of the plastic zone (Figure 4.7). Portion of the crack situated in the plastic zone experiences a constant, negative pressure of the same magnitude as the yield stress, σ_y. Plastic zone dimension, ρ, is such that the stress singularity should disappear; thus, the superposition of SIF due to remotely applied stress, K_σ, and that due to plastic zone closure, K_ρ, should be zero, giving:

$$K_\sigma + K_\rho = K = 0 \tag{4.61}$$

Solving Eq. (4.61) using both K_σ and K_ρ in terms of ℓ and ρ, a final expression is obtained as

$$\frac{\ell}{\ell + \rho} = \cos\left(\frac{\pi \sigma_\infty}{2\sigma_y}\right) \tag{4.62}$$

At very low remote stress levels, σ_∞, $\sigma_\infty \ll \sigma_y$, the stress expansion of right-hand side of 4.62 leads to

$$\rho = \frac{\pi}{8}\left(\frac{K}{\sigma_y}\right)^2 \tag{4.63}$$

Barenblatt reached the same solution assuming cohesive forces closing the virtual crack due to a plastic zone. Simplified Dugdale/Barenblatt model 96

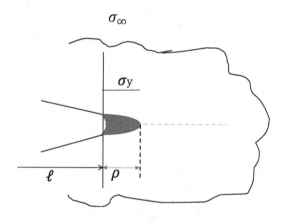

Figure 4.7 Constant negative pressure σ_y on the extended plastic zone size ρ in front of a crack of length ℓ under external stress σ_∞.

is almost equivalent to Irwin model (4.60) [Parker, A.P. (1981)]. All these models considered plastic zone shape using oversimplified yield criterion. Newer models considered much advanced yield criterion like Von Mises (1913) and were able to predict the plastic zone through the thickness

$$\left(r_p = \frac{K_I^2}{2\pi\sigma_y^2} \cos^2\frac{\theta}{2}(1-2v)^2 + 3\sin^2\frac{\theta}{2} \right)$$ suitable for metals [Parker, A.P. (1981)].

Such models show that plastic zone size is a function of v and angular location from crack tip. When stress triaxiality disappears $(v = 0)$, the plastic zone size reduces to something similar to (4.63) at $\theta = 0$. For a rubbery material when $v = 0.5$, plastic zone disappears (which is supported by data) at $\theta = 0$.

Experiments have shown that this zone of plasticity is centered at the crack tip. Aforementioned equations give the approximate ideal radius of the plastic zone deformation beyond the crack tip, which is useful to many structural scientists because it gives a good estimate of how the material behaves when subjected to stress. In the aforementioned equation, the parameters of the stress intensity factor and indicator of material toughness, K_{1c}, and the yield stress, σ_y, are of importance because they illustrate many things about the material and its properties, as well as about the plastic zone size. For example, if K_{1c} is high, then it can be deduced that the material is tough, and if σ_y is low, one knows that the material is more ductile. The ratio of these two parameters is important to the radius of the plastic zone. For instance, if σ_y is small, then the squared ratio of K_{1c} to a σ_y is large, which results in a larger plastic radius. This implies that the material can plastically deform, and, therefore, is tough. This estimate of the size of the plastic zone beyond the crack tip can then be used to more accurately analyze how a material will behave in the presence of a crack.

The J integral has enjoyed great success as a fracture characterizing parameter for nonlinear materials by Rice, J.R. (1968) idealizing elastic–plastic deformation as nonlinear elastic. This provided the basis for extending Fracture mechanics methodology well beyond the validity limits of LEFM. The loading behavior for the two materials is identical, but the material responses differ when each is unloaded. It was proven mathematically that the J integral is equivalent to the energy release rate in nonlinear elastic materials. The J integral represents a way to calculate the strain energy release rate, or work (energy) per unit fracture surface area, in a material. The theoretical concept of J integral was discussed earlier, but in 1968, James R. Rice showed that an energetic contour path integral (called J) was independent of the path around a crack. Thermodynamic origin for J-integral is explained in last few sections in this chapter.

Experimental methods were developed using the integral that allowed the measurement of critical fracture properties in sample sizes that are too small for Linear Elastic Fracture Mechanics (LEFM) to be valid. These experiments allow the determination of fracture toughness from the critical value of fracture energy J_{1c}, which defines the point at which large-scale plastic yielding during propagation takes place under mode I loading. In the mid-1960s, James R. Rice (then at Brown University) and G.P. Cherepanov independently developed a new toughness measure to describe the case where there is sufficient crack-tip deformation so that the part no longer obeys the linear-elastic approximation. Rice's analysis, which assumes nonlinear elastic (or monotonic plastic deformation theory) deformation ahead of the crack tip, is designated as the J integral. This analysis is limited to situations where plastic deformation at the crack tip does not extend to the furthest edge of the loaded part. It also demands that the assumed nonlinear elastic behavior of the material is a reasonable approximation in shape and magnitude to the real material's load response. The elastic–plastic failure parameter is designated J_{1c} and is conventionally converted to K_{Ic} using the equation in the next sentence. Since engineers became accustomed to using K_{Ic} to characterize fracture toughness, a relation has been used to reduce J_{1c} to it: $K_{Ic}^2 / E = J_{1c}$, for plane stress and $J_{1c} = \dfrac{K_{Ic}^2}{E}\left(1 - \vartheta^2\right)$, for plane strain. Also note that the J integral approach reduces to the Griffith theory for linear-elastic behavior.

J integral is equal to the strain energy release rate for a crack in a body subjected to monotonic loading. This is generally true, under quasistatic conditions, only for linear-elastic materials. For materials that experience small-scale yielding at the crack tip, J can be used to compute the energy release rate under special circumstances such as monotonic loading in mode III (antiplane shear). The strain energy release rate can also be computed from J for pure power-law hardening plastic materials that undergo small-scale yielding at the crack tip. The quantity J is not path-independent for monotonic mode I and mode II loading of elastic–plastic materials, so only a contour very close to the crack tip gives the energy release rate. Also, Rice showed that J is path-independent in plastic materials when there is no nonproportional loading. Unloading is a special case of this, but nonproportional plastic loading also invalidates the path-independence. Such nonproportional loading is the reason for the path-dependence for the in-plane loading modes on elastic–plastic materials. For isotropic, perfectly brittle, linear elastic materials, the J integral can be directly related to the fracture toughness if the crack extends straight ahead with respect to its original orientation.

R-curve was an early attempt in the direction of elastic–plastic Fracture mechanics, also known as Irwin's crack extension resistance curve or crack growth resistance curve. This curve acknowledges the fact that the resistance to fracture increases with growing crack size in elastic–plastic materials. The R-curve is a plot of the total energy dissipation rate as a function of the crack size and can be used to examine the processes of slow stable crack growth and unstable fracture. R-curve in Fracture mechanics was an attempt to quantify the principle of increasing fracture resistance as a function of crack length suitable for materials with crack tip plasticity. Many materials with high toughness do not fail catastrophically at a particular value of J or CTOD. Rather, these materials display a rising R-curve, where J and CTOD increase with crack growth. However, the R-curve was not widely used in applications until the early 1970s. The main reasons appear to be that the R-curve depends on the geometry of the specimen and that the crack driving force may be difficult to calculate.

In metals, a rising R-curve is normally associated with the growth and coalescence of micro-voids. In the initial stages of deformation, the R-curve is nearly vertical; there is a small amount of apparent crack growth due to blunting. As J increases, the material at the crack tip fails locally and the crack advances further. Because the R curve is rising, the initial crack growth is usually stable, but instability can be encountered later. One measure of fracture toughness J_{1c} is defined near the initiation of stable crack growth. The precise point at which crack growth begins is usually ill-defined. Consequently, the definition of J_{1c} is somewhat arbitrary. For plane strain state, fracture occurs with very little crack tip plasticity, and fracture resistance R is constant. The R-curve relates to the length-dependent resistance of a fracture and to propagation in terms of fracture energy [Broek, D. (1974), Parker, A.P. (1981)]. It is founded on the principle that failure occurs when the energy release rate G approaches the crack resistance R. A typical R-curve involves plotting G as a function of crack length and plotting the R values on the same plot as a function of crack length (Figure 4.8a and b) for crack growth [Anderson, T.L.]. As shown earlier, G can be calculated as $G = \frac{\pi}{E}\sigma_\infty^2 \ell$ for plane stress and $G = \frac{\pi}{E}\sigma_\infty^2 \ell (1 - v^2)$ for plane strain as a function of crack length. R is calculated from actual crack growth experiments taking the crack tip plasticity into account for each incremental growth. Once G equals R, catastrophic failure occurs at a value of G_c. Shape of R-curve depends on material behavior and, to a lesser extent, on the geometry of the cracked structure. R-curve for a brittle material is flat because the surface energy is an invariant property [Griffith, A.A. (1921)]. R-curve can take a variety of

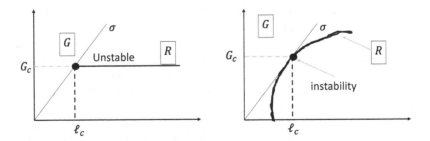

Figure 4.8 (a) Flat R-curve—Griffith instability and (b) rising R-curve.

shapes in materials with nonlinear load response and/or associated with plastic deformation ahead of the crack tip.

> The worst sin in an engineering material is not lack of strength or lack of stiffness, desirable as these properties are, but lack of toughness, that is to say, lack of resistance to the propagation of cracks.
>   Before one can start arguing about how tough materials ought to be, one must be able to measure how tough they are.
>
> J.E. Gordon
> (The new Science of Strong Materials or Why You Don't
> Fall though the Floor)

In summary, the progress from Galileo's days to modern times followed a general trend where mechanics advanced considerably, but the physics of failure criterion and lifetime prediction made very little progress. We first introduced strength as stress, strain, or energy criterion. Quite often, strain energy density (area under the stress-strain curve) for a brittle material was referred to as toughness in early days. In ductile materials, the area swept by stress–strain curve contains both stored and absorbed energy, and the combined energy was loosely used as toughness. Once it was realized that all fracture initiates from a preexisting defect, resistance to crack propagation had to be included into the definition of toughness. Thus, strength is reflected in stress, strain, and energy criteria while toughness is expressed through stress intensity factor, SIF (1962), crack opening displacement (1963–64), and energy release rate criteria.

Lifetime Prediction—Empirical and Physics-Based Approaches

One main topic mysteriously absent in all this treatment is *time to failure* (lifetime) till entropy density criterion made its debut [Chudnovsky, A.I. (1973)]. It is quite natural since mechanics does not explicitly deal with time. This gap was then addressed through phenomenological kinetic equations of crack propagation where K, G, and J were used as the driving force. Lack of a good foundation resulted in innumerable kinetic equations depending on specific material and loading condition [Benham, P.P. and Warnock, F.V. (1973), Paris, P.C. and Erdogan, F. (1963), Knott, J.F. (1973)]. The first attempt to address time to failure in engineering structures began with a concept that most failures are accelerated due to non-monotonic load applications. Thus, *regressing* to strength concept in early strength of materials, engineers tried to determine number of cycles (N) to failure on application of a cyclic stress range (S) for manufactured materials. This was introduced as S–N curve [Benham, P.P., and Warnock, F.V.] even though Fracture mechanics had already been introduced. S–N approach does not separate out crack initiation and propagation stages and is certainly inapplicable to structures with preexisting defects. Paris and Erdogan were first to introduce a fatigue crack propagation law, bringing SIF into play analyzing a large number of fatigue test data [Paris, P.C. and Erdogan, F. (1963)]. For SIF range $\Delta K \left(K_{max} - K_{min} \right)$, crack growth rate $\partial \ell / \partial N$ is expressed as

$$\partial \ell / \partial N = C \left(\Delta K \right)^m \tag{4.64}$$

where C and m are experimentally determined material constants (adjustable parameters). Although this and many variations of this empirical equation have gotten acceptance, they fail to address the crack-initiation side of failure. In simple situations, where crack initiates quickly and follows fairly rectilinear propagation, Eq. (4.63) has some utility. Note that Eq. (4.63) is normally called Paris' law. Incidentally, Paul Paris was a student of Erdogan in Lehigh University. Crack propagation is commonly studied under constant load (such as creep) as well as under variable load (such as fatigue). Specimen geometry and loading conditions in most of the experimental studies are selected in the way that assures a rectilinear crack path and mode I conditions, so that K is the only SIF. The fatigue loading conditions have many characteristics. For example, periodic loads are characterized by the maximum and minimum stresses σ_{max} and σ_{min}, stress range $\Delta \sigma = \sigma_{max} - \sigma_{min}$, stress ratio $R = \dfrac{\sigma_{max}}{\sigma_{min}}$, and frequency and shape function of load variation within a cycle; more parameters are needed in case the fatigue load variation is a random process. SIFs variation follow that of the remote load since SIF is a linear function of the load. Thus, there are K_{max} and K_{min}; and the ratio $R = \dfrac{K_{max}}{K_{min}}$. Apparently, creep is a special case of fatigue when $R = 1$. Thus, a formulation of the creep crack growth equation is straightforward for these kinds of approach as there should be a functional relation between the crack growth rate $\dot{a} \, vs \, K$ (*authors of this book chooses to use ℓ for crack length elsewhere in this book but Paris et al. preferred "a"*). For fatigue conditions, the problem is more complex since it involves more parameters. At the same time, fracture under fatigue loading is quite important from practical viewpoint: Cracks grow much faster and therefore the lifetime of structures is much shorter under fatigue, as compared to creep conditions (at the same maximum load). A typical crack growth behavior in double logarithm coordinates $\left(\log \partial a / \partial t \, versus \, K \right)$ is shown in Figure 4.9. There is a so-called short crack stage usually associated with crack initiation. *A separate section will be dedicated in this book to this topic of crack initiation which constitutes most of the lifetime of any structure. This topic is skipped over by most*

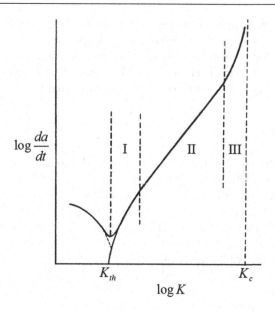

Figure 4.9 Typical fatigue crack growth behavior.

researchers or mentioned in passing. Under fatigue conditions, the crack initiation is commonly associated with formation and growth of surface defects, whereas under creep conditions, a crack may originate both inside of a structural component and on its surface. Examples of crack initiation in the bulk material (engineering thermoplastics) under creep condition were given by Chudnovsky et al. (2012). The location of the initiation site may strongly affect the duration of the slow crack growth stage. The time associated with crack initiation may vary widely from 20% to 80% of the total lifetime. In contrast with a poorly documented crack initiation, the slow crack growth stage is usually a well observable and reproducible process that readily allows modeling. This is one of the reasons why most works on brittle fracture are focused on slow crack growth. Since it is difficult to determine SIF for such short cracks, this stage, strictly speaking, does not belong to Figure 4.9. However, it is a part of the incubation period in the process of macroscopic crack formation, and, as such, it contributes to the duration of the fracture process with no detected crack propagation.

Thus, there is a threshold of SIF range ΔK_{th}, below which no macroscopic crack growth takes place. On another end, there is a critical value ΔK_c at which SCG suddenly changes to rapid crack propagation. The interval between $\log \Delta K_c$ and $\log \Delta K_{th}$ can be divided into three subintervals corresponding to different stages of fatigue crack growth, as indicated in Figure 4.9. The first stage follows the crack initiation and corresponds to a very low crack growth rate, but a relatively high acceleration. The second stage is characterized by almost linear relations between $\log \frac{\partial a}{\partial t}$ and K; the power law of Paris and Erdogan applies within this stage. At the end of that stage, fast crack acceleration starts and a critical state is approached, where a transition to RCP takes place. Thus, the Paris–Erdogan equation is not applicable for stages I and III, though it is a good approximation of the second stage. However, stage III is relatively short and does not significantly affect the lifetime. Regarding stage I, Eq. (4.63) overestimates the crack growth rate in that stage and thus gives a conservative estimate of lifetime. Overall, a simple empirical equation is a reasonably good approximation for observed fatigue crack growth behavior in stage II. The surprising feature of Eq. (4.63) is that K_{max} does not enter the equation explicitly, as one may intuitively expect. Another puzzle is that Eq. (4.63) does not give any meaningful prediction for creep crack growth, when $R \rightarrow 1$, $\Delta K \rightarrow 0$, although creep loading is a special case of fatigue as claimed. Other limitation of Eq. (4.63) is that it does not capture the actual crack growth behavior near the threshold ΔK_{th} (before) and within stage I as well as approaching ΔK_c (within stage III). Only stage II is described well by Eq. (4.63). Several modifications of Eq. (4.63) have been proposed to eliminate the limitations. Most of the propositions have a certain level of generality, and are discussed later in the chapter. Predicting SCG portion of the curve by empirical equation is not an improvement over S-N approach. At least, S-N approach addresses total lifetime, and, with advanced statistical technique, one can provide information on reliability of structures.

A modification of Eq. (4.63) that models the crack behavior within stage III was introduced by Forman et al. (1967). This equation improves the Paris Law in two ways: (a) It accounts for crack behavior at Stage III, and (2) it extends Paris Law, by incorporating the R-ratio. Stage III improvement is achieved by adding one adjustable parameter, i.e., K_c, that is commonly considered as a toughness parameter. On approaching K_c, crack growth rate approaches ∞, a transition to rapid crack propagation. However, it was found later that the value of K_c is not a material parameter but rather depends on the loading history and crack growth process. A modification that covers all three stages was proposed by Erdogan and Ratwani (1970) by introducing six empirical constants. This equation indeed captures all three stages of crack growth depicted in Figure 4.9. However, this is done by introducing four additional adjustable parameters, in comparison with the original Paris–Erdogan model. This results in a dramatic drop of the predictive power of the model due to uncertainty in evaluating large number of adjustable parameters in view of the inevitable scatter of fatigue data. Others proposed modifications of Paris–Erdogan equation tailored to fit experimental data for materials and loading configurations they studied [Walker, K. (1970)] and is of limited value for other materials and test conditions. It was pointed out by Elber that general limitations of Paris Law are associated with irreversible deformation near crack tip, causing a temporary retardation of crack growth under certain conditions [Elber, W. (1970)]. More general limitations of Paris Law are associated with irreversible deformation near crack tip. In metals, it is commonly observed as a plastic deformation zone that has a strong effect on crack growth. Elber (1970) observed that crack closure prior to full unloading is reached, as it would have happened in a perfectly elastic material. The stress at which the closure took place at unloading part of the loading cycle was denoted by σ_{clos}. There was also a crack opening stress σ_{op} observed on the subsequent active part of loading cycle. Elber (1970) suggested that crack closure was caused by an irreversible deformation behind crack tip. Localized large deformation prior to fracture resulted in a local contact of the deformed material prior to the rest of fracture surfaces coming into contact. In experiments, the crack opening (closure) was not uniform along the crack front as well as in the crack growth directions. Thus, an experimental determination of closure and opening stresses was a challenging task. However, the employment of high-resolution microscopy combined with experiments on relatively soft materials with large plastic deformation zone (aluminum) provided convincing evidence for the Elber suggestion. In a later work, Elber (1971) reported that crack propagation took place only after the applied stress σ overcomes the opening stress σ_{op}.

Thus, it is not the stress range that controls fatigue crack growth, as suggested by Paris Law. Phenomenological modeling of the crack growth retardation was proposed by Wheeler (1972). This model works very well for constant stress amplitude loading and occasional overloading. However, a generalization for an arbitrary variation of the stress amplitude proposed by Wheeler (1972) requires additional adjustable parameters and thus is of limited value. Several further modifications of Paris Law aimed at a better fit of specific experimental data existing. Most of them deal with specific materials and/or test conditions. A critical introspection of all such developments has been presented in more details elsewhere [Chudnovsky, A., *International Journal of Engineering Science* (2014)]. More fundamental approach requires an analysis of the crack driving force which will be elaborated by the authors in this book.

However, it is generally accepted that failure is a process which begins virtually at the instant of load application to a specimen. From the point of view of molecular structure, the process of failure can be regarded as the development of structural irregularities, i.e., breakdown of the initial order. No structure, man-made or natural, is created defect-free. Thus, the theory of elasticity based on homogeneity of space and time is wrong from the beginning and can predict fracture phenomena as long as the defects do not become active. But fracture is a critical phenomenon associated with activating internal defects in a structure that apparently looks like continuum. The process is thus intimately related to changes of the system entropy. Thermodynamically, the similarity of the melting and failure processes lies in this order to disorder transition. One cannot address quantitative estimation of lifetime of engineering structures (structural reliability, i.e., probability of "no failure" for specified duration) without closely examining the root causes of failure and understanding fracture mechanisms.

In 1980s, several classical strength of materials researchers like Ashby recognized the importance of the identifications of the specific fracture mechanisms associated with the mode of fracture and the origin of failure in form of Fracture Mechanism Maps. These maps usually consist of some form of stress plotted against some kind of temperature axis, typically stress normalized using the modulus versus homologous temperature with contours of strain rate [Ashby, M.F. and Tomkins, B. (1980)]. Professor Ashby was the pioneer in such kind of studies in ceramics and metals. Few such diagrams are shown in Figures 4.10 and 4.11 with classifications from observations.

To generate such a map, one needed to first classify various types of fracture mechanisms such as "brittle" cleavage versus ductile intergranular creep and brittle fracture, ductile fracture, and trans-granular fracture. These types of maps required a huge body of research on each specific material. At this point, one can summarize everything we have discussed so far on the classical approach to fracture by stating that there are too many strength criteria and that no universal one can emerge out of this complexity. Furthermore, there is no time in Classical Strength Criteria, and *entropy density criterion* was the

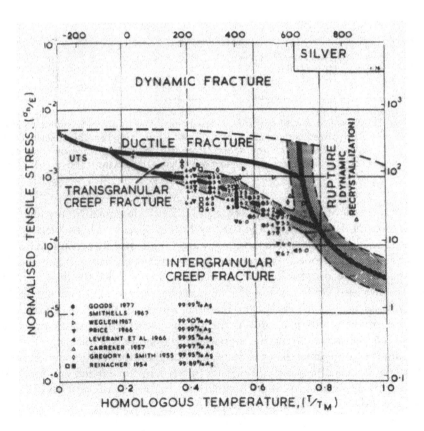

Figure 4.10 Fracture Mechanisms Map in silver.

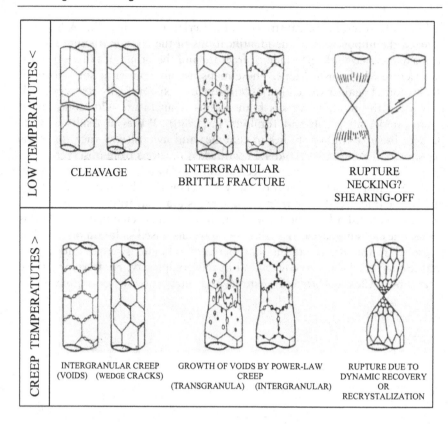

Figure 4.11 Classification of Fracture Mechanisms: The upper row refers to low normalized temperature (<0.3 T_m is the melting temperature), where plastic flow does not depend strongly on temperature or time; the lower row refers to a relatively high temperature above 0.3 T_m in which materials creep.

first to introduce time [Chudnovsky, A.I. (1973)] Consequently, any lifetime prediction model is material and loading history specific. Microdefects play a decisive role in fracture. However, they are not adequately accounted for in classical strength models. Fracture mechanism maps present a practical tool for material selection for durable applications, but it is not the final answer either. They also did not deal with fractography and the stochastic nature of fracture surface, which is the true material witness of the sequence of dramatic events that took place during the advancement of the fracture process. There is no Universal Strength Criterion, since fracture is associated with the activation of different type of defects depending on the loading conditions. However, we need to recognize that thermodynamic instability condition is universally applicable to different situations, but this consideration did not gain grounds in classical approaches.

At this point, we need to introduce thermodynamics to develop a complete framework for physics of fracture as proposed by current author in this book. Fracture processes consist of nucleation and growth of microdefects. For thermodynamic description of fracture, one needs to introduce a list of parameters of state by incorporating a damage parameter. L.M. Kachanov was the first one who did it explicitly. Since then, numerous papers suggesting various damage parameters and constitutive equations for them have been published. Summaries of some of the proposed damage models can be found in Kachanov, L.M. (1958), Kachanov and Montagut (1986) and Krajcinovic, D. et al. (1981). Specific interpretation of a damage is vitally important for establishing a correspondence between experimental studies and a damage model. It is not so important for general thermodynamics.

In thermodynamics, the *Helmholtz free energy* is a thermodynamic potential that measures the useful work obtainable from a closed (this assumption is true in planetary scale, but "closed" has no meaning in universe scale) thermodynamic system at a constant temperature and volume (isothermal, isochoric). Furthermore, at constant temperature, the Helmholtz free energy is minimized at equilibrium. The concept of free energy was developed by Hermann von Helmholtz, a German physicist, and first presented in 1882. In contrast, the Gibbs free energy or free enthalpy is most commonly used as a measure of thermodynamic potential (especially in chemistry) when it is convenient for applications that occur at constant *pressure*. The concept of Gibbs free energy was developed in the 1870s by the American scientist Josiah Willard Gibbs.

The Helmholtz energy is defined as follows [Adams, A.W. (1979)]:

$$F_\varepsilon = U - TS \tag{4.65}$$

where, F_ε is the Helmholtz free energy (SI unit: joules, CGS unit: ergs); U is the internal energy of the system (SI: joules, CGS: ergs); T is the absolute temperature (in Kelvin) of the surroundings; S is the entropy of the system (SI: joules per Kelvin, CGS: ergs per Kelvin).

Quality of Energy

Second law of thermodynamics is primarily about quality of energy. *Energy quality* is a measure of the ease at which a form of energy can be converted to useful work or to another form of energy. A high-quality form of energy is easily converted to work or to a lower quality form of energy, whereas converting low quality forms of energy to work or a higher quality form may be inefficient, difficult, or impossible. Low quality energy is any form of energy which is dispersed and disorderly and has less potential or ability to be utilized for work. From quality point of view, electromagnetic and/or electrical energy is the highest quality energy followed by mechanical, chemical, and heat. Quality of energy leads to another definition of entropy. As the entropy of a closed system naturally increases, the energy quality decreases. First law of thermodynamics is about the conservation of energy where energy can only change forms, but the total energy in a closed system remains constant. This leads to the following famous relation.

$$dU = \delta Q \pm \delta W \tag{4.66}$$

If work is done on the system, "+" sign is used, and if the work is done by the system "−" sign is used for changes in work. Going back to (4.64) and differentiating results in

$$\delta U = \delta F_\varepsilon + T\delta S \tag{4.67}$$

Equation (4.66) applies to an isothermal process since temperature is constant. At constant T, change in entropy can be expressed as (utilizing Eqs. 4.65 and 4.66)

$$\delta S = \frac{\delta W - \delta F_\varepsilon + \delta Q}{T} \tag{4.68}$$

For work done on the system, Helmholtz free energy can be related to Gibb's potential energy G as

$$G = F_\varepsilon - W \tag{4.69}$$

When no work is done on the system, Gibb's energy is the same as Helmholtz free energy. $\delta S \geq 0$ and $\delta G \leq 0$ in real processes. When S maximizes, G minimizes. These conservation laws applied to theory of elasticity lead to the famous J integral presented next.

$$-\delta G_k = J \tag{4.70}$$

Few other useful expressions in elasticity can be expressed in thermodynamic framework by introducing the concept of free energy density $f(\tilde{\varepsilon}, T)$ and $g(\tilde{\sigma}, T)$. Taking a volume integral of $f(\tilde{\varepsilon}, T)$ for a body under stresses and strains, we generate Helmholtz free energy. Similarly, taking a volume integral of $g(\tilde{\sigma}, T)$ for such a body, we generate Gibb's potential energy. This results in thermodynamic definitions of stress and strain expressed in the following equation pair:

$$\tilde{\sigma} = \frac{\partial f(\tilde{\varepsilon}, T)}{\partial \tilde{\varepsilon}} \text{ and } -\tilde{\varepsilon} = \frac{\partial g(\tilde{\sigma}, T)}{\partial \tilde{\sigma}} \tag{4.71}$$

Thermodynamic relationship encompassing work done through stress $(\tilde{\sigma})$ and strain $(\tilde{\varepsilon})$ can be written as

$$g(\tilde{\sigma}, T) = f(\tilde{\varepsilon}, T) - \tilde{\sigma} : \tilde{\varepsilon} \tag{4.72}$$

At this point, we need to introduce Gauss's theorem from vector calculus. *Gauss' theorem* [Katz, Victor J. (1979)] is a theorem which relates the flux of a vector field through a closed surface to the divergence of the field in the volume enclosed (Figure 4.12). In other words, g is a function of σ and T, which in turn depends on position x. But g is not explicitly dependent on x. This logic reduces to $\partial_k g = 0$ for homogeneity assuming homogeneous temperature field.

$$\int_V div\bar{A}(x)dV = \int_{dV}\bar{A}.\bar{n}dA \tag{4.73}$$

where $\bar{n}$ represents the normal vector. Other than conservation laws in the theory of elasticity, one needs to take following things into consideration such as no body forces expressed as $\sigma_{ij,j} = 0$. For a plane problem, this can be written as

$$\sigma_{11,1} + \sigma_{12,2} = 0 \tag{4.74}$$

$$\sigma_{21,1} + \sigma_{22,2} = 0$$

Strain components can be written as derivative of displacement vectors u_i. This set of equations is expressed next.

$$\varepsilon_{ij} = \frac{1}{2}\left(u_{i,j} + u_{j,i}\right)$$
$$\varepsilon_{xx} = \frac{1}{2}\left(u_{x,x} + u_{x,x}\right) \tag{4.75}$$
$$\varepsilon_{xy} = \frac{1}{2}\left(u_{x,y} + u_{y,x}\right)$$

where $u_{y,x}$ stands for $\dfrac{\partial u_y}{\partial x}$. Next thing to consider is that temperature, T, is constant for isothermal process (application of stress-strain field). Also, conservation of impulse must be maintained as well. Impulse conservation means the product of the force applied and the time over which it is applied remains the same before and after the collision. Impulse in physics is a term that is

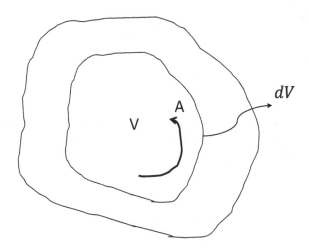

Figure 4.12 Volume element V whose boundary is dV, with the arrow representing area A.

used to describe or quantify the effect of force acting over time to change the momentum of an object. Knowing that total energy, $U = K + P$ where K and P stand for kinetic and potential energies, we introduce *Hamiltonian* action function, also called Hamiltonian as $A = K - P$. A is a mathematical definition introduced in 1835 by Sir William Rowan Hamilton to express the rate of change in time of the condition of a dynamic physical system. Thus, for such systems, A must be minimized. Last but not the least requirement is homogeneity of elastic medium. In homogeneous media, the elastic constants and density ρ are constant everywhere. Very few materials can be considered as homogeneous elastic medium. For small deformation, real medias can be approximated as homogeneous.

Calculation of the Potential Energy Release Rates

One can construct four types of potential energy release rates for a hole, crack, discontinuities, or a body of discontinuities translating, rotating, expanding, and distorting within an elastic field. The following transformation can be applied to the elastic potential energy density $\rho(g,T,P)$ using arbitrary operator "δ" and assuming homogeneous temperature field.

$$\frac{\partial \rho}{\partial P} \delta P = \delta \rho - \frac{\partial \rho}{\partial \tilde{\sigma}} : \delta \tilde{\sigma} \qquad (4.76)$$

Therefore,

$$\int_{V_A} \frac{\partial \rho}{\partial P} \delta P dV = \int_{V_A} \left[\delta \rho - \frac{\partial \rho}{\partial \tilde{\sigma}} : \delta \tilde{\sigma} \right] dV \qquad (4.77)$$

Now one can use (4.71) for various operators.
(a) Using the operator of an infinitesimal translation δ_k^{tr} in kth direction, one can modify (4.71) as

$$-\int_{V_A} \frac{\partial \rho}{\partial P} \delta_k^{tr} P dV = \int_{V_A} \left[\delta_k \rho - \frac{\partial \rho}{\partial \tilde{\sigma}} : \delta_k \tilde{\sigma} \right] dV \qquad (4.78)$$

The left hand side of the equation is $-J_k$, where $k = 1, 2$. Incorporating relationship in indices form where f is the elastic strain energy, equations of equilibrium in the absence of mass forces are $\partial_j \sigma_{ij} = 0$.

$$\rho = f(\tilde{\varepsilon}, T, P) - \sigma_{ij} : \varepsilon_{ij}^e \qquad (4.79)$$

The expression within the bracket on right hand side of 4.78 can be simplified as

$$\partial_k f - \partial_j (\sigma_{ij} u_{i,k}) \qquad (4.80)$$

The expression within brackets in Eq. (4.79) can be simplified as

$$\partial_k \rho \cdot \frac{\partial \rho}{\partial \sigma_{ij}} \partial_k \sigma_{ij} = \partial_k \left(f - \partial_j \left(\sigma_{ij} \varepsilon_{ij}^{(e)} \right) \right) + \varepsilon_{ij}^{(e)} \partial_k \sigma_{ij} = \partial_k f$$

$$- \partial_k \left(\sigma_{ij} \varepsilon_{ij}^{(e)} \right) + \partial_k \left(\sigma_{ij} \varepsilon_{ij}^{(e)} \right) - \sigma_{ij} u_{i,jk} = \partial_k f - \partial_j \left(\sigma_{ij} u_{i,k} \right)$$

(4.81)

Substituting 4.80 into 4.77 and using Green's theorem, we find that for $k = 1$, J_1 stands for conventionally used energy release rate with respect to crack extension perpendicular to the direction of applied stress.

$$J_1 = \int_{\partial V_A} \left[f n_1 - \sigma_{ij} n_j u_{i,1} \right] d\Gamma$$

(4.82)

where Γ is a contour surrounding the crack tip, starting from the lower crack surface, and moving anticlockwise to the upper surface. Strain energy in direction 1 is $f n_1$, and $\sigma_{ij} n_j$ stands for the traction vector applied outward on contour Γ. In contemporary Fracture mechanics, J_1 is expressed as energy release rate (ERR).

(b) Using the operator of an infinitesimal rotation, we obtain

$$\delta_m^{rot} = \varepsilon_{klm} x_k \partial_l + \left(\begin{array}{c} \textit{Correction for tensor components} \\ \textit{associated with rotation of coordinate} \\ \textit{system spinorial terms} \end{array} \right)$$

(4.83)

where ε_{klm} is the alternating symbol with components equal to +1 or −1 for even and odd permutations of 1, 2, 3, respectively, and 0 for all other combinations. Substituting 4.80 into 4.77 and providing transformation similar to that in 4.81, one can obtain:

$$\int_{V_A} \frac{\partial \rho}{\partial P} \delta_m^{rot} P dV = L_m$$

(4.84)

Here, L is a pseudovector ($m = 3$ for 2D problems).

Simplifying the term under the integral in Eq 4.83 using 4.73, L_m can be rewritten as

$$L_m = \varepsilon_{klm} \int_{\partial V_a} \left[n_k x_y f - x_y \sigma_{ij} n_j u_{i,k} - \sigma_{yi} n_i u_k \right] d\Gamma$$

(4.85)

For an isotropic medium, the path independency of L_m remains similar to that of J_k for homogeneous medium. Consequently, the partially arbitrary contour (see Figure 4.13) $\Gamma = \Gamma^t u \Gamma^*$ can be used instead of ∂V_a in 4.84.

(c) The operator of an infinitesimal isotropic expansion is

$$\delta^{exp} = -x_k \partial_k$$

(4.86)

Substituting 4.85 into right hand side of Eq. (4.84), we obtain

$$\int_{V_A} \frac{\partial \rho}{\partial P} \delta^{exp} P dV = M \tag{4.87}$$

where

$$M = \int_{\partial V_a} \left[n_k x_k f - \sigma_{ij} n_j u_{1,k} \right] d\Gamma \tag{4.88}$$

The M-integral (4.87) possesses the path-invariancy for linear medium only. The J_k, L_m, and M integrals represent the potential energy release rates with respect to the translation, rotation, and isotropic expansion of any kind of discontinuity or discontinuities in an elastic medium due to application of a stress field. In a similar fashion, a second rank tensor N_{kl} can be introduced as the potential energy release due to deviatoric deformation. N_{kl} is particularly sensitive to deviations in the stress and strain fields near the crack tip. In contrast to J_k and L_m, M and N_{kl} do not possess path independency (see *Appendix II* for historical remarks). Notably, the path independency of J_k, L_m, and M renders their evaluation more convenient although it has no physical significance [Eshelby, J.D. (1951), Rice, J.R. (1968)]. For instance, for nonhomogeneous, anisotropic, and nonlinear medium, the Eqs. (4.81), (4.84), and (4.87) express the same potential energy rates as discussed before, but all the integrals are not path independent. We will soon find out that new fracture theories (Crack layer theory) will effectively incorporate these fundamental thermodynamic expressions. Other than J-integral, most of these integrals are missing in contemporary Fracture mechanics.

At this point, we need to introduce some new terminology for crack propagation in materials slightly deviating from conventional nomenclature. If crack propagation is possible in any material, we would like to coin the failure process as brittle fracture, otherwise the failure is ductile. Ductile failure occurs by large-scale yielding or flow-type process associated with crack blunting. Thus, there is no need for Fracture mechanics treatments in later situations as attempted by many researchers for polymer thin films. Brittle failure can then be classified into two major types such as *Cooperative Brittle Fracture* and *Solo Brittle Fracture*. We will finally bring both fracture types under the same umbrella of statistical Fracture mechanics (*SFM*) in the concluding section.

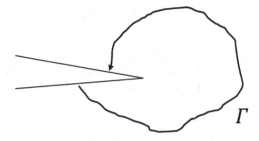

Figure 4.13 A crack and a contour Γ around the crack.

Chapter 5

Cooperative Brittle Fracture (Lifetime Prediction)

As summarized in previous chapter, just examining the fracture mechanisms is not adequate. Failure is a process which begins virtually at the instant of load application to a specimen. If we compare failure with death of a human being, Indian scriptures say that the death process of a human body begins at the instant a life is born. Thus, our life's journey appears to be from birth to death through the unavoidable process of aging in cellular and sub-cellular level. Similarly, submicroscopic and microscopic damage accumulation under a stress/strain field (incubation process) leads to a macroscopic crack nucleation or initiation. Let us call time to nucleation as t_n. Once nucleated, a stable crack propagation occurs for a period of time t_p, and finally crack becomes unstable and SCG gets transformed into a dynamic (fast) process leading to avalanche-like crack propagation (t_a). For all practical purpose, $t_a = 0$. Most of the Fracture mechanics studies address crack propagation in notched specimen, thus, t_n reduces to 0→30% of the total lifetime and propagation comprises the rest. For an unnotched specimen, the relation is almost inverse, i.e., $t_n \gg t_p$, but the nucleation time is less deterministic. The location of the initiation site may strongly affect the duration of the SCG stage. The time associated with crack initiation may vary widely from 20% to 80% of the total lifetime t_f. In contrast with a poorly documented crack initiation, the SCG stage is usually a well-observable and reproducible process that readily allows modeling. This is one of the reasons why the majority of works on brittle fracture are focused on SCG. This book will address the topic of crack initiation in detail in the following section and propose approaches to allow statistical predictions.

Fracture initiation (FI) is the least studied and the most uncertain stage of fracture phenomena. It is the end result of a microscopic-scale damage accumulation process, rather than an instantaneous event. Using a micro-heterogeneous material (sandstone), initiation studies were conducted using the standard indirect tensile strength test [American Society for Testing and Materials (ASTM 1989) and The International Society for Rock Mechanics (ISRM 1978)]. Acoustic emissions, optical microscopy, and scanning electron microscopy (SEM) are employed for monitoring and characterizing

DOI: 10.1201/9781003359845-6

the discrete micro-mechanical events preceding macroscopical fracture. The observations suggest that brittle fracture initiation is the end result of a microscopic damage accumulation process. In a material that can develop a process zone, fracture initiation will be followed by a period of slow crack propagation (SCG) till catastrophic failure ensues. In a material like sandstone, fracture initiation after the incubation period of damage accumulation will lead to final failure. Thus, sandstone is an ideal material for studying incubation or nucleation period t_n leading to fracture initiation. In such a material, $t_p = t_a = 0$. Thus, $t_f = t_n$. In the following paragraph, we will discuss a model for estimating t_n. The approach should be applicable to cooperative fracture in any material if we can identify the micro damage elements in the specific material.

A simple statistical model of micro damage accumulation leading to brittle fracture in a micro-heterogeneous material is also proposed [Jasarevic et al. (2009)]. For this purpose of testing, the standard "Indirect tensile strength test", also known as the Brazilian test [(*ASTM* (*1989*) and *ISRM* (*1978*)], was selected. In this test, a disc is compressed by point forces along the diameter of the disc. The analytical solution for elastic stress distribution in such a setting is available (Chen et al. (1998)). It includes the expression for the maximum tensile stress in terms of the magnitude of the applied force and dimensions of the disc. The effect of rock anisotropy, i.e., layering orientation with respect to the loading direction, is also addressed. In the Brazilian test, the fracture initiation is commonly observed in the central region of the disc, where the tensile stresses are maximal. To detect the processes that accompany fracture initiation in micro-heterogeneous materials such as rocks, acoustic emission (AE) monitoring techniques have been applied for many decades (Hoagland et al. (1975), Suzuki et al. (1980), Hashida and Takahashi (1993). In this work, Torry Buff sandstone was used for an illustrative purpose, and it displays a typical brittle behavior. The Torry Buff sandstone is a very fine-grained porous rock (porosity ~19%, permeability ~3-millidarcy, Young's modulus 10.5 MPa measured on three 44.5-mm diameter, 92.4-mm length cylinders, and dry unconfined compressive strength ~40 MPa). Our observations support the findings of Hashida and Takahashi (1993) after employing AE cumulative energy parameter to detect and quantify the damage accumulation prior to FI. In addition to AE monitoring, the loose material fragments were collected from the fracture surfaces, and statistical analysis of the fragment sizes based on the transmission optical microscopy observations was performed. The fragments vary in size from a fraction of an individual grain to a large cluster of grains. The structure of fragments has been observed using scanning electron microscopy (SEM). The presence of many fragments is a direct evidence of multiple micro-fracture events leading to macroscopic FI, in agreement with the AE data. AE energy and the average fragment size can be linked by the energy analysis of fragmentation [Glenn et al., 1986].

The original publication [Jasarevic et al. (2009)] displays a thin section micrograph of Torry Buff morphology observed using transmission polarized light microscopy. It reveals the grain size, shape and orientation, grain clusters, grain boundaries, and pores. The morphological micro-heterogeneity can be characterized, for example, by the normalized standard deviation (SD) of the grain cluster (GC) size averaged over a set of sampling windows of the same size at different locations. The variation of the SD of the average GC size with the sampling window size shows that the SD decreases with increasing sampling window size. An RV was defined with respect to morphological heterogeneity as the minimal volume size, above which the SD doesn't change appreciably. The RV for the Torry Buff is about 0.7–1 mm. Thus, starting from such an RV, the micro-heterogeneous structure of Torry Buff can be presented as a homogeneous continuum with certain properties such as elastic constants, permeability, and other volume-averaged characteristics obtained as the average over RV. However, it should be noted that this RV size may not be sufficient for strength properties due to well-known scale (or size) effect on such properties (see, e.g., Tetsuya et al. (2005)).

There are deterministic and probabilistic approaches to modeling FI in micro-heterogeneous materials. A number of deterministic models stem from Continuum Damage Mechanics initiated by L.M. Kachanov (1958). The basic attributes of such models are: (a) a damage parameter (a scalar, second rank tensor, or more complex object) in addition to the conventional set of parameters of state such as stress, strain, and temperature; (b) a kinetic equation for the damage parameter evolution; and (c) a condition of FI: FI is assumed to be associated with either a critical value of damage parameter (a fracture criterion) or, in a more interesting version, an unbounded damage growth rate, i.e., a material instability with respect to damage growth (see details in Kanuan and Chudnovsky (1999)). There are also two different probabilistic approaches to modeling FI. One is based on nucleation models of statistical thermodynamics (Golubovic and Feng (1991), Berdichevsky and Khanh (2005)). This method is well justified for thermoactivated processes when the nucleus size may increase and decrease at random in accordance with a Markovian process. However, it is not obvious that the same arguments are applicable to rock-type materials with a preexisting population of microcrack defects of various sizes and locations, which is not associated with a Markovian random process. The second statistical approach uses an analogy between the first-order phase transition and fracture, which is based on crystal instability considerations originally proposed by M. Born (1939, 1940). Advances in computational power made it possible to numerically analyze the analogy between fracture and a phase transition [Buchel, A. and Sethna, J.P. (1996)]: a representative volume (RV) of a micro-heterogeneous solid is explicitly treated as a system of connected discrete elements with random strength. When one of the elements breaks, the stress acting on that element is transferred to a number of surviving neighboring elements.

Straightforward analysis shows that when a certain critical number of elements break, the entire RV fails, which is interpreted as the FI event on a macro scale. That approach follows the classical Ising model in ferromagnetism [Ising, E. (1925)]. Various models of this kind differ in the stress transfer rule as well as in the choice of probability distribution for the individual element strength. For example, Curtin (1998) assumes that the stress is transferred as the inverse cube of the distance between the surviving and broken elements. Shicker and Pfuff (2006) used a simplified stress transfer rule that involves only the four nearest neighboring elements. The modeling part of the present work follows a nearest-neighbor stress redistribution rule and adopts the Weibull distribution for the individual element strengths.

Jasarevic et al. (2009) used 50.8-mm diameter and 25.4-mm thickness ($t/D = 0.5$) discs cut from a block of Torry Buff sandstone for all tests. Compression of the discs is performed with curved steel jaws following the ASTM (1989) and ISRM (1978) standards. The image of the central domain of the specimen shows a disconnected fracture path. To soften the steel–rock contacts, the discs are wrapped by a double layer of paper masking tape. The Brazilian tests are conducted on a servo-hydraulic loading system using a crosshead speed of 0.01 mm/sec. The average critical stress measured is 0.364 N / mm^2, which is computed as $\sigma_{cr} = -(2W_f) / (\pi D \delta)$ (Chen et al. (1998), where W_f is the peak load recorded at fracture, and D δ are the disc diameter and thickness, respectively. The standard deviation of the peak stress was 0.015 N / mm^2. A video recording of one face of each disc was made during each test with the objective of observing the onset and growth of the macroscopic fracture. An AE sensor was attached to the other face of the disc near the center and AE data collected using an AE monitoring system. Two characteristics, the number of AE events and the instantaneous AE energy, were recorded. The total kinetic energy is emitted via acoustic waves within a certain frequency range.

In all tests, the discs fractured along the disc diameter coinciding with the loading direction. FI took place at the disc central region as anticipated. The speed of the video recording was not sufficient to monitor in detail the post-FI fracture growth. A few frames were captured that reveal a few disconnected fractures in the center that quickly coalesce and grow to split the sample. AE monitored during the loading directly indicates the time-dependent damage accumulation process prior to the macroscopic FI. This damage accumulation is also manifested in multiple fragmentations. A cumulative damage (CD) parameter $\omega(t)$ can be used as the number of new surfaces per unit volume formed from the beginning of loading up to a current time t as a product of fragment volume and number of fragments $(\omega(t) = \pi a^2(t).N(t))$, where $a(t)$ stands for an average fragment diameter formed at the time $0 < t \le t$ and $N(t)$ is the total number of fragments formed from the beginning of loading up to the time t. For a relatively small strain rate, the average fragment diameter is directly related to the stress level σ at the time of fragment formation [Glenn et al. (1986)], Young's modulus (E), and fracture surface

energy $\gamma \left(a(t) \approx \dfrac{2E\gamma}{\sigma^2(t)} \right)$.

Based on Glenn–Chudnovsky model, the large fragments are formed at a low stress, and the smallest fragments are formed at the onset of the macroscopic fracture under high stress. The criterion for onset of macroscopic FI was set as $d\omega / dt\rho \to \infty$. A simple statistical model for simulating the observed FI process was formulated. Consider an elementary material volume ΔV consisting of $n3$ microelements, e.g., grains. All the elements of ΔV have the same elastic properties but differ in toughness represented by a random specific fracture energy (SFE). The elementary volume ΔV fails, if $n2$ elements of any 2D cross section of ΔV perpendicular to the maximal tensile stress are broken. It represents the fracture initiation at a point of a continuum. Each of $n2$ elements is bearing the stress until the break point. The following energy failure criterion for an individual microelement with a characteristic size a and SFE γ was adopted assuming that E represents the elastic modulus at any microelement within the continuum:

$$\sigma^2 a / 2E \geq 2\gamma \tag{5.1}$$

SFE γ in the right-hand side is assumed to be a random field with mutually independent values at each of $n2$ elements that obey a three-parameter Weibull distribution [Weibull, W. (1951)]. A video of the damage process leading to fracture initiation is in Video 1 (see Support Material: www.routledge.com/9781032418131, and Video Captures 5.1–5.5). The fracture in sandstone is an ideal model for fracture initiation in plastics where initiation results from the accumulation of fracture events in micro scale.

Slow Crack Growth (SCG)

Returning to SCG, we already introduced one popular empirical kinetic equation for crack propagation (Paris–Erdogan) which in certain situations could be useful but fails measurably in most others. We have seen that many empirical equations for SCG are applicable only in a relatively narrow range of conditions and cannot characterize more general crack growth scenarios. To understand these limitations, Chudnovsky et al. developed the concept of a crack driving force (a thermodynamic force [*Appendix IV*]), crack equilibrium, and stability conditions using thermodynamics considerations. The work by Chudnovsky et al. focuses on the physical aspects of fracture, more specifically on the SCG process. After analyzing the limitations of prevailing approaches to the problem, an alternative approach associated with the crack-layer concept that is based on understanding of the underlying mechanisms of SCG will be presented. In a stable crack-loading configuration, when the energy release rate is a decreasing function of crack length, the crack growth equation follows from the crack equilibrium condition assuming the crack resistance is constant. It allows a straightforward experimental examination. Experiments demonstrate that the theoretical prediction of stable crack growth is in substantial disagreement with observations. In particular, the theory does not account for the formation of microdefects that significantly affect the crack behavior. In an unstable crack-loading configuration (the energy release rate is an increasing function of crack length), the second law of thermodynamics prohibits crack growth in most elastic solids with constant specific fracture energy 2γ (Griffith surface energy). Such conclusion conflicts with numerous observations of SCG in unstable configurations. The conflict is caused by the formation of distributed damage around the crack that absorbs energy and leads to an increase of 2γ with crack propagation.

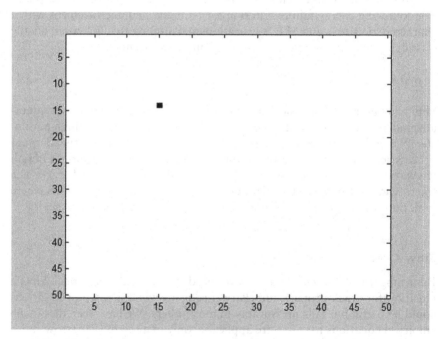

Video Captures 5.1–5.5 Screenshots of computer simulation of a crack-initiation event as the end result of a relatively long process of sub-microscopic damage accumulation process. Each frame displays several damage formation events (represented by black boxes). Once a chain of black boxes combine, crack initiation takes place.

Note: The crack initiation is the most mysterious and the least addressed topic in literature; it represents the beginning of the second stage of the fracture process, that is, the slow crack growth stage. This video is based on the Ph.D. thesis work of Dr. Haris Jaserevich (Ph.D. graduate student at UIC under the supervision of Prof. A. Chudnovsky).

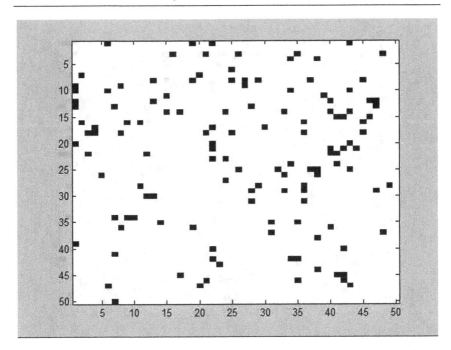

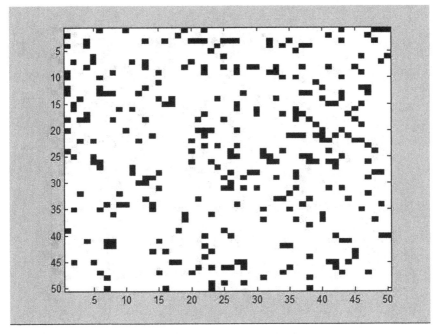

Video Captures 5.1–5.5 (Continued)

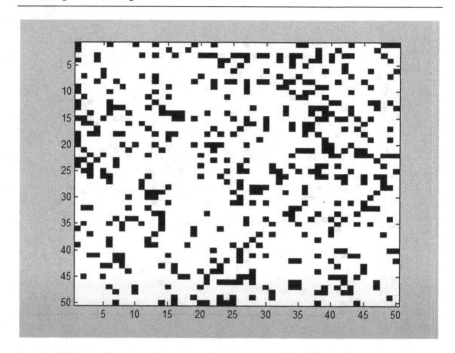

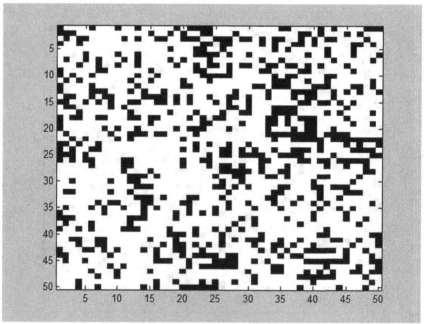

Video Captures 5.1–5.5 (Continued)

Let us call the entire damage zone including the crack as *Crack layer* (CL). Then, we discuss coupling between distributed damage and a crack that constitutes CL. Experiments have been conducted on a transparent thermoplastic that allows optical microscopy studies in the process of fatigue testing. The crack–damage interaction affects the thermodynamics of the process. The CL driving forces and kinetic equations of CL evolution follow from more detailed thermodynamic analysis. The effect of CL on slow crack growth resistance (commonly known as fracture toughness) illustrates the consequences of distributed damage evolution. Experimental examination of toughness variation and a simple explanation of the R-curve behavior are discussed. It is shown that CL poses several degrees of freedom thus leading to a system of coupled constitutive equations. CL theory also established the laws governing the SCG and estimating the time prior to transition to rapid growth is of crucial importance. Earlier propositions based on stress concentration coefficients lose meaning when the singular elastic stress field at the tip of an ideally sharp smooth crack has been recognized. In most of the engineering materials, a real crack is never ideally sharp, and the crack trajectory is never smooth. It is common to use a homogeneous continuum model by most researchers, where physical properties of each point are identified with the properties averaged over an RV. The RV size may be different for different properties, and RV may not exist for the properties exhibiting macroscopic scale effect (see RV discussion in Zhang et al. (2013)). Modeling of any features within RV calls for micromechanical considerations. A macroscopic modeling of crack commonly considers smooth crack trajectories and ideally sharp crack tips. For a linear elastic material, this implies the well-known square root singularity of stresses: $\sigma_{ij}^{I,II,III}(r,\theta) = K_{I,II,III} f_{ij}^{I,II,III}(\theta)/\sqrt{\pi r}$ for each three modes of loading, I, II, and III. Here r and θ stand for distance from the crack tip and angular position of any point with respect to the crack tip in an infinite linear elastic medium. The modeling of a crack by an ideal cut in an elastic solid is well justified by the energy considerations. Indeed, the computation of elastic energy release rate G_1 associated with crack advance using the square root singularity at the crack tip provides a very good approximation of more detailed modeling involving blunted crack tip and small-scale yielding [Slepjan, L.I. (1990)]. Moreover, the energy release rate G_1 is simply related with SIFs: $G_1 = \dfrac{K_I^2 + K_{II}^2}{E} + \dfrac{K_{III}^2}{2\vartheta}$ where E and ϑ stands for modulus and Poisson's ratio. Note the subscript "1" in G_1 stands for the direction of crack extension (x1 direction of a 3D coordinate system) and is not related to mode I (crack opening mode) stress state. The expression of energy release rate provides a physical meaning of SIFs. Under fatigue conditions (cyclic loading), the crack initiation is commonly associated with formation and growth of surface defects, whereas under creep conditions, a crack may originate both inside of a structural component and on its surface.

Examples of crack initiation in the bulk material (engineering thermoplastics) under creep condition were given by Chudnovsky et al. (2012). One essential thing to note is that crack is always accompanied by damage which some addresses through different models of plasticity. However, such models are not suitable for most materials, especially plastics.

Most plastics have very localized deformation events (or discontinuities) like crazes and shear bands that comprise the damage zone. *Micromechanics* was developed to find solutions for such damage zone comprising individual micro-damages or discontinuities. The key point of micromechanics of materials is the localization, which aims at evaluating the local (stress and strain) fields in the phases for given macroscopic load states, phase properties, and phase geometries. Such knowledge is especially important in understanding and describing material damage and failure. Because most heterogeneous materials show a statistical rather than a deterministic arrangement of the constituents, the methods of micromechanics are typically based on the concept of the representative volume element (RV). An RV is understood to be a subvolume of an inhomogeneous medium that is of sufficient size for providing all geometrical information necessary for obtaining an appropriate homogenized behavior. Most methods in micromechanics of materials are based on continuum mechanics rather than on atomistic approaches such as nano-mechanics or molecular dynamics. All these approaches may be subsumed under the name of "continuum micromechanics". Several scientists tried to address this topic, including current authors.

Alan Lesser addressed micromechanics of crack–microcrack interactions by incorporating law of superposition and a double layer potential method in his PhD thesis (Alexander Chudnovsky served as a co-advisor) [Lesser, 1989]. In this study, localized discontinuity resulted from the formation of densification bands as the authors of that work called them the densification that most likely resulted from the fact that the experiments were conducted under background of the hydrostatic compression, in contrast with the crazing observed in (PS), under uniaxial tension; see the work by J. Botsis at al. The micromechanics portion of the analysis in Alan's thesis was restricted to a case of an infinite plate made of a linear elastic material. General problem of crack–microcrack interaction was decomposed into three separate problems: (a) a uniform plate free of all cracks and microdefects subjected to a uniform shear and compressive stress, (b) a plate with the main crack only and a traction equal to the shear stress in (a) applied to the crack surface, and finally (c) a plate with both main crack and the array of microcracks with displacements applied along the array (Figure 5.1).

Double layer potential technique to calculate the displacement field and thereby the stress field was borrowed from previous work by Chudnovsky et al. (1987a), M.; [Chudnovsky, A. and Kachanov, M. (1983)]. Interaction between a gliding dislocation and a crack was evaluated using 3D Buckner–Rice weight function theory [Rice, J.R. and Thompson, R. (1974)]. Ballarini et al.

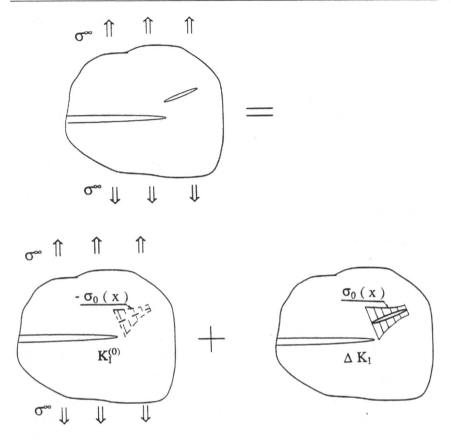

Figure 5.1 Superposition of the crack tip fields.

addressed the problem of interaction between a crack and a dislocation dipole where they present the results of an analysis which considers the interaction between a semi-infinite crack and a dislocation dipole [Ballarini, R., Denda, M. (1988)]. Applying the operator derived by Ballarini and Denda to the crack/dislocation interaction solution developed by K.K. Lo (1978), explicit expressions are obtained for the stress intensity factors at the tip of the crack. Results are computed and discussed for a variety of geometrical configurations, with the intent of developing an understanding of the effects of position and orientation of the dislocation dipole on crack tip shielding and anti-shielding. According to Ballarini et al. the solution can also be used as a Green's function in semi-empirical analyses such as the one proposed by Chudnovsky et al. (1987a) and Chudnovsky and M. Kachanov (1983), where the interaction between a crack in a polymeric material and the damage (crazing) which surrounds it is solved by experimentally measuring the opening displacements of the crazes and calculating the amount of toughening caused by the damage. The problem of the interaction of a crack and a dislocation in a medium with a nonlinear stress–strain law is considered for the case of a semi-infinite crack in a displacement-loaded strip under longitudinal shear deformation by Atkinson (1966). A power law of stress–strain relation is considered, and the dislocation is assumed positioned so that the net effect of its interaction with the crack is to produce a zero-stress intensity factor when combined with the effect of the applied displacements. Thus, the Atkinson–Kay super dislocation model of a relaxed crack tip is extended to a specific situation where the material satisfies a power-law stress–strain relationship. A much more accurate approach applicable to crack and damage zone interaction in CL will be discussed next. Instead of assumptions about the damage zone contribution to crack as proposed by Irwin and others in Fracture mechanics, one can actually identify the damage zone elements (that varies from material to material) and use micromechanics for a semi-empirical estimate.

Micromechanics of CL

Damage zone ahead of a macroscopic crack in micromechanics can be treated as the interaction of main crack with an array of microcracks. Although mathematics is developed for traction-free microcrack faces, it can be easily extended to crazes in polymers where the displacement is delimited by the craze fibrils spanning across the surface. In the case of shear banding or slip bands in polymers and metals, local displacements due to the bands replace displacements at the microcrack. As stated, earlier micromechanics of crack–micro-crack interactions can be solved by incorporating the law of superposition (Figure 5.1) and a double layer potential method. In order to translate displacement at a point in space to another point, we need to introduce second Green's tensor of elasticity. Figure 5.2 illustrates the problem and the method schematically. Green's functions are objects of fundamental importance in field theories, since they provide the fundamental solution of linear inhomogeneous partial differential equations (PDEs) from which any particular solution can be obtained via convolution with the source term. In

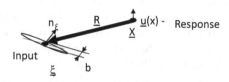

Figure 5.2 A microcrack opening displacement "b" is the input, and the response $u(x)$ occurs at a position X at a distance R. n_x is the normal at position X.

the context of elasticity, the Green's function of rank two is also known as the Green's tensor.

Strain field operator is expressed in the equations appearing after Eq. (5.2):

$$\tilde{\Phi}\left(\overline{\xi},\overline{n}_{\xi}\left(\varphi,\theta\right);\overline{x}\right) = \frac{1+\vartheta}{2\pi R^{2}} \cdot \left\{\overline{R} \otimes \overline{R}\right\}$$

$$\tilde{\Phi} = \left\{\frac{1-\vartheta}{1+\vartheta}\left[\overline{n}_{\xi R} \otimes \overline{R} - \overline{R} \otimes \overline{n}_{\xi} + \left(\overline{n}_{\xi} \cdot \overline{R}\right)I\right] - 2\frac{\overline{n} \cdot \overline{R}}{R^{2}}\overline{R} \otimes \overline{R}\right\} \tag{5.2}$$

Strain tensor can be presented as a gradient of the displacement vector $\overline{u}$ and finally related to the stress as a product of Young's modulus and the strain tensor.

$$\tilde{\varepsilon} = \left(\overline{\nabla}\overline{u} + \overline{u}\overline{\nabla}\right); \quad \tilde{\sigma} = \tilde{E}:\tilde{\varepsilon} \tag{5.3}$$

Stress tensor at position x can be described as given here where F is the stress operator.

$$\tilde{\sigma}\left(x\right) = \int_{-a}^{a}\tilde{b}\left(\xi\right) \cdot \tilde{F}(\xi,\overline{n}_{\xi},\overline{x})d\xi \tag{5.4}$$

Stress field operator at ξ, x is described here where the second term on left hand side describes the gradient of strain field operator and E is the Young's modulus.

$$\tilde{F}_{x}(\overline{\xi},\overline{n}_{\xi},\overline{x}) = \overline{E}:\left(\overline{\nabla}_{x}\tilde{\phi}\left(\overline{\xi},\overline{n}_{\xi},\overline{x}\right)\right)_{s} \tag{5.5}$$

Crack layer can be simplified as a traction-free crack of length L surrounded by numerous arrays of parallel microcracks as shown in Figure 5.3. Next, considering a crack surrounded by an active zone ahead of its tip essentially modelled as an array of microcracks, one can arrive at the following set of useful relationships: (a) main crack and remote load and (b) for the CL-active zone modelled as microcrack array. Once again, using superposition principle, one can derive the following:

Main crack and remote load:

a) $-\overline{n}_{x} \cdot \int_{-L}^{0}\tilde{\delta}\left(\overline{\xi}\right) \cdot \tilde{F}(\overline{\xi},\overline{n}_{\xi},\overline{x})d\xi + \tilde{\sigma}^{0}\left(x\right) \cdot \overline{n}_{x} + \overline{n}_{x} \cdot \int_{V}\rho\left(\overline{\xi}\right)\left\langle\overline{b}\left(\xi\right)\right\rangle \cdot F\left(\overline{\xi},\overline{n}_{\xi},\overline{x}\right)dV = 0 \quad x \in \left(-L,0\right)$

Microcrack array:

b) $\overline{n}_{x} \cdot \int_{-L}^{0}\tilde{\delta}\left(\overline{\xi}\right) \cdot \tilde{F}(\overline{\xi},\overline{n}_{\xi},\overline{x})d\xi + \overline{n}_{x} \cdot \tilde{\sigma}^{0}\left(x\right) \cdot \overline{n}_{x} + \overline{n}_{x} \cdot \int_{V}\rho\left(\overline{\xi}\right)\left\langle\overline{b}\left(\xi\right)\right\rangle \cdot F\left(\overline{\xi},\overline{n}_{\xi},\overline{x}\right)dV\left[1 - \delta\left(\overline{\xi} - \overline{x}\right)\right]dV = 0$

$$\text{(5.6a and b)}$$

δ-function has been used to combine crack, array of microcracks, and a space free of any crack or microcracks under the influence of remote load.

The formulation of the problem requires basic equations such as equilibrium equation, Hooke's law, small strain consideration, and compatibility condition.

$$div\,\tilde{\sigma} = 0 \quad \text{Equilibrium}$$

$$\tilde{\varepsilon} = \tilde{C} : \tilde{\sigma} \quad \text{Hook's Law}$$

$$\tilde{\varepsilon} = \left(\overline{\nabla u}\right)_s \quad \text{Small Strain} \tag{5.7}$$

$$Rot\,\tilde{\varepsilon} = 0 \quad \text{Compatibility}$$

General solution for the displacement due to crack and microcrack interactions can be written as Eq. (5.8):

$$\bar{u}^{tot}(\bar{x}) = \bar{u}^0(x) + \int_{-L}^{0} \tilde{\delta}(\xi) \cdot \tilde{\phi}(\xi, n_\xi, \bar{x}) d\xi + \int_V [\int\int_{(s)} \rho(\xi; \varphi, \theta) \langle \bar{b} \rangle (\xi) \cdot \tilde{\phi}(\bar{\xi}, n_\xi(\varphi, \theta); \bar{x}) d\varphi d\theta] dV \tag{5.8}$$

Average COD of microcracks with orientation (j,q) within elementary RV dV at the point x within the microcrack array is shown in Figures 5.2 and 5.4. Using the aforementioned equations, one can plot Green's function for SIF due to microcrack (unit discontinuity dipole) surrounding the crack tip under both mode I and mode II stress state. As an example, Figures 5.5 and 5.6 are presented here where the dipole is oriented at $0°$ and $20°$ with x axis.

Finally, using the aforementioned equations and superposition method, one can describe the crack damage (process) zone interaction for a crack of length L (Figure 5.7). The arrays of curves emanating from the crack tip forward and backward present contours of equal level of the mode I SIF Green's function $G_1^{SIF}(\xi_i)$ due to unit normal displacement dipole with $\overline{n(0,1)}$ and $b_2(0,1)$ at the point ξ_i [Ben Ouezdou, M. and Chudnovsky, A. (1988) Chudnovsky et al. (1987a); Wu, S. and Chudnovsky, A. (1991)]. The continuous lines indicate SIF amplification effect due to the discontinuities in front of the crack; whereas the two "butterfly wings" of dotted lines show the reduction of SIF, i.e., shielding effect of the discontinuities located on both sides of the crack. By distributing the discrete crazes and passing to a continuum damage, craze density is introduced as the bivector $\rho(\xi_i)$ that represents the oriented area of the middle craze plane per unit volume (with dimensions $\left(\rho = mm^2/mm^3\right)$). In addition, the total SIF (see Figure 5.7) is the sum of SIF K_1^0 due to remote load and the increment of SIF due to distributed crazing

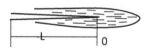

Figure 5.3 Crack of length L surrounded by an array of parallel microcracks. O is the center of the crack tip.

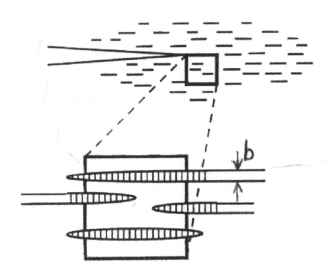

Figure 5.4 Picture of crack and the microcrack array. The rectangular element is the representative volume d*V* expanded below showing microcrack COD of "*b*".

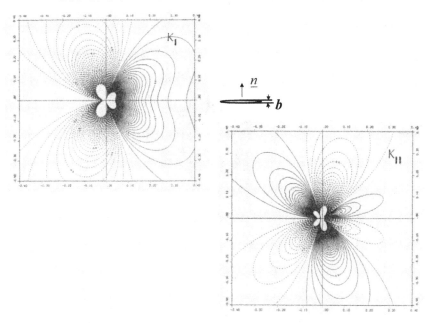

Figure 5.5 Plot of nondimensionalized Green's Function for SIF due to unit discontinuity (dipole) causing mode-I and mode-II stress state at the crack tip. The dipole is oriented at 0° with *x* axis.

with opening $\tilde{b}$ dependent on the crack–craze array interaction [Ben Ouez-dou, M. and Chudnovsky, A. (1988), Chudnovsky et al. (1987b), Wu, S. and Chudnovsky, A. (1991)]. The calculations conducted for the case shown in Figure 5.7 (right upper section) suggest that the shielding effect (a negative SIF increment) is stronger than the amplification and the total SIF is vanishing [Chudnovsky et al. (1987b)]. It results in the "beak" type crack tip-opening displacement profile that can be noticed in the Figure 5.7 [Chudnovsky, A. (2007)]. The aforementioned example indicates that there is a domain of PZ, where the defects constituting PZ strongly interact with crack by shielding and/or amplifying the crack tip fields. The crack in turn causes the new defects formation in that domain. Such a domain is referred to as the active zone (AZ) of CL. The diagram in Figure 5.7 clearly establishes the amplification zone where microcracks continue to grow and shielded zone where microcracks are unloaded. The amplification zone is also referred to as active

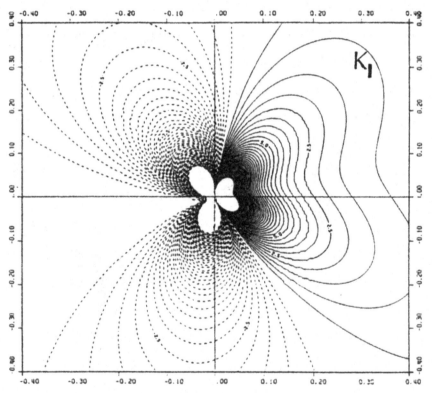

Figure 5.6 Plot of nondimensionalized Green's Function for SIF due to unit discontinuity (dipole) causing mode-I and mode-II stress state at the crack tip. The dipole is oriented at 20° with x axis.

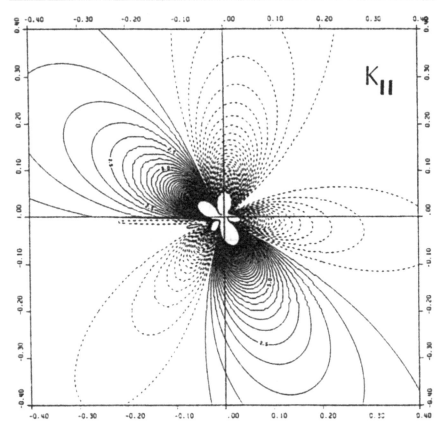

Figure 5.6 (Continued)

zone and the shielded zone as wake zone (from similarity with the wake of a travelling ship or a boat). The equation for total SIF K_1^{tot} is shown in the bottom of the diagram (Figure 5.7). The calculations conducted for the case shown in Figure 5.7 suggest that the shielding effect (a negative SIF increment) is stronger than the amplification, and the total SIF is vanishing [Chudnovsky et al. (1987b)]. In order to apply these concepts to the experimental work done by Botsis et al. (1988), one can extract the actual measurement of average displacement due to crazes inside an RV within the damage zone (Figure 5.8). A plot of displacement due to the entire CL can then be easily calculated and plotted as a function of normalized crack length (Figure 5.9). Now the micromechanics model can be used to predict the crack growth rate combining with CL theory discussed in the next section. Figure 5.10 shows these predictions and agreement with actual data obtained from experiments in PS.

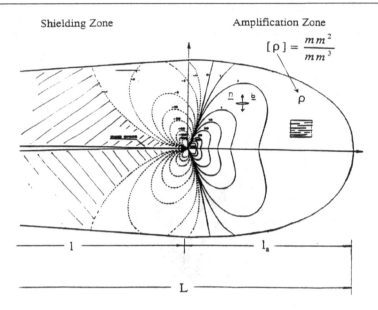

$$K_1^{tot} = K_1^0 + \int_{V_A} G_1^{SIF}(\xi) b_2(\xi) \rho(\xi) dV$$

Figure 5.7 Crack–damage (consisting of microcracks) zone interactions; expression for total SIF using superposition.

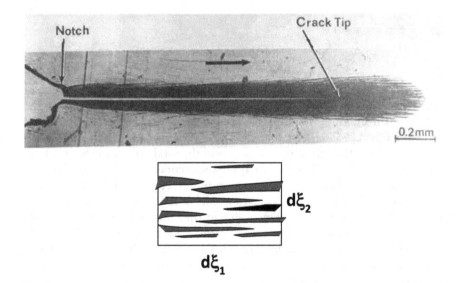

Figure 5.8 CL in polystyrene (PS) and experimental measurement of average displacement in the crazes inside a representative volume within the damage zone.

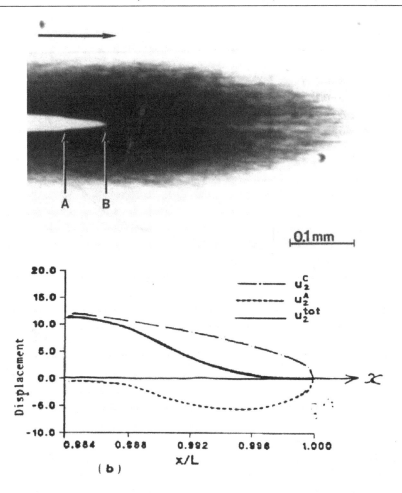

Figure 5.9 Picture of CL and a plot of displacement due to CL as a function of distance normalized with crack length.

Figure 5.11a illustrates a CL formed under fatigue loading in polystyrene (PS), an amorphous brittle thermoplastic. It is a composite view of side and cross-sectional views of the CL damage zone. Figure 5.11a and b shows optical micrographs of the side view of PZ (dark area) that surrounds the open crack. Bottom section on the left hand side of Figure 5.11b displays the cross section of the lower part of PZ perpendicular to crack 1 near the crack tip for two fractured specimens indicated by two arrows. The cross section reveals an array of crazes well aligned with the crack faces. Individual crazes are well visible as horizontal dark lines at the peripheral domain of PZ. The stepwise profile of the cross section of the fracture surface suggests that it was formed by micro-fracturing of individual crazes developed in front of the crack tip before

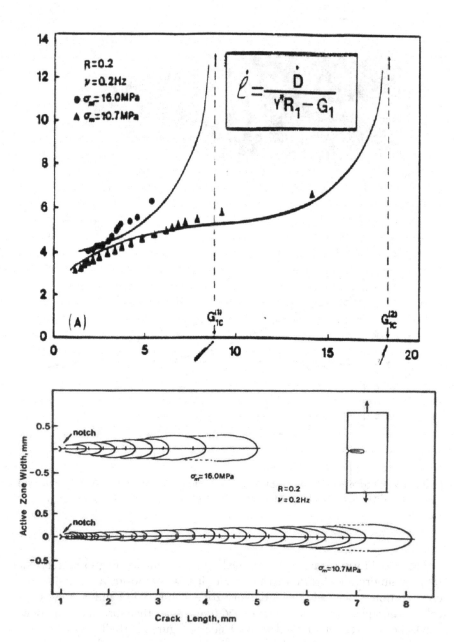

Figure 5.10 Prediction of crack growth rate for two different loading histories in PS compared with actual data. Figure on the bottom shows the evolution of CL damage zone.

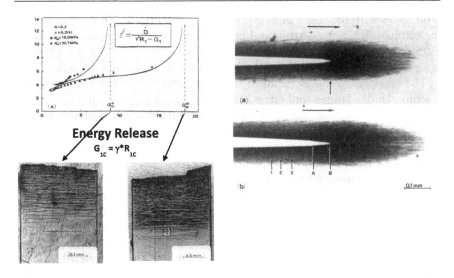

Figure 5.11 Side view and the cross-sectional views of the CL damage zone.

the growing crack has reached that location. There is a strong interaction between crazes in a vicinity of the crack front and the crack [Ben Ouezdou, M., and Chudnovsky, A. (1988), Chudnovsky et al. (1987b), Chudnovsky, A. (2007)]. It results in significant modification of the crack tip fields.

Considering the two types of damage zone associated with CL propagation, one could examine apparent toughness. For the same material PS, one finds two different G_{1c} values depending on loading condition proving that the fracture toughness is *not a material parameter*. However, using the relationship between G_{1c} and $\gamma^* R_{1c}$, one notices that the difference can be attributed to resistance parts $R_{1c}^{(1)}$ and $R_{1c}^{(2)}$ reflecting the extent of damage in each case. Botsis et al. showed that the specific energy of damage γ^* is constant in PS which appears to be a material parameter. It makes physical sense since γ^* is associated with craze formation in PS, a more fundamental parameter associated with molecular characteristics of the polymer [Botsis, J., March 1988, June 1989; Botsis et al., January 1983; Kramer, E.J. (1984)]. This results in a very simple expression for the ratio of critical energy release rate shown as:

$$G_{1c}^{(1)} / G_{1c}^{(2)} = R_{1c}^{(1)} / R_{1c}^{(2)} \tag{5.9}$$

There are commonly recognized three major stages of cooperative brittle fracture: (a) Crack initiation; (b) slow crack growth (SCG); (c) crack instability and transition to either rapid crack propagation (RCP) or ductile rupture. Later in the chapter, we briefly discuss the initiation stage and focus our attention on SCG. The third stage occurs very fast, since the rapid crack propagates with the speed of a few hundred meters per second or faster, and the ultimate fracture of structural components by RCP takes a fraction of second. It means the lifetime of plastic component consists of incubation time prior to crack initiation and the duration of stable slow crack growth.

The crack initiation is the most uncertain and the least studied stage of fracture. Under fatigue and creep loading conditions, crack initiation is the conclusion of a damage accumulation process on sub-micro and microscales as discussed earlier. This accumulation process may be called a fracture incubation stage (FIS), since it does not manifest itself on a macro scale commonly used in continuum mechanics. FIS progresses faster in the vicinity of preexisting micro-defects such as cavities, gel, and/or foreign particles or agglomerates. Figure 5.12 shows at 100× magnification, a domain of fracture surface containing the site of crack initiation from a foreign particle. The particle is located above the center of two concentric rings surrounding the initiation site. The rings are formed in discontinued crack growth process that follows a continuous growth of a small crack after its initiation. The fracture surface shown in Figure 5.12 was formed in an accelerated test at an

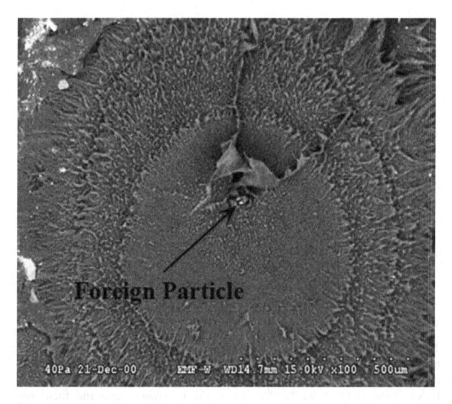

Figure 5.12 Fracture initiation site in polyethylene pipe at a foreign particle (×100).

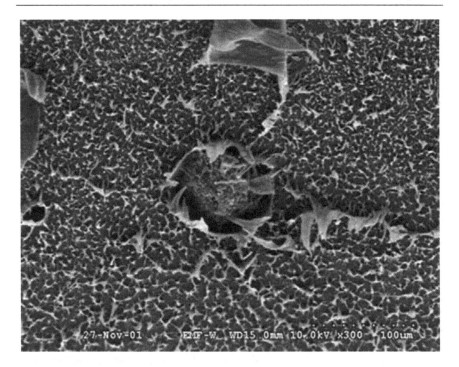

Figure 5.13 A catalyst housing fragment at the center of the crack-initiation site.

elevated temperature. The duration of FIS strongly depends on service or test conditions (stress level, strain rate, and temperature) as well as on the chemical composition of the defect, its shape, size, location as well as the type of interface with resin matrix. Depending on specific conditions, the duration of FIS may constitute between approximately 20% and 80% of the total lifetime of plastic component. A large scatter commonly observed in the lifetime of plastic part is primarily associated with random geometry and location of the particle triggering the crack initiation. For example, under the same loading conditions, for the particle of the same shape and size, fracture initiation at the middle wall of PE pipe results in almost an order of magnitude longer life than that for fracture initiation near the inner surface [Choi et al. (2007)].

Figure 5.13 presents a peculiar case of crack initiation from a micro fragment of the housing of a catalyst that controls polymerization process and thus is responsible for polymer formation in the first place. The ceramic catalyst-housing fragment is visible at the center of the micrograph (300× magnification). A set of small and large fibers drawn from the base PE surrounds and partially covers the fragment. Thus, the crack initiation that leads to the end of the useful life of a particular polymer was caused by a particle directly related to polymerization process at the beginning of the useful life.

In contrast with the scenario described earlier, the crack initiation in degradation–driven brittle fracture does not need a defect to start the process. The nature of fracture initiation in such cases is different. The surface of plastic part apparently is in direct contact with the surrounding (usually aggressive environment). Therefore, a thin surface layer degrades faster than the interior portion of the part. Chemical degradation of polymers is manifested in the reduction of molecular weight (MW), i.e., a measure of the length of polymer chain. For semi-crystalline polymer like PE, the reduction of MW often leads to an increase of crystallinity and therefore material density. The densification of surface layer in amorphous polymers results from physical aging. The densification of surface layer attached to interior-unchanged material results in a buildup of residual stresses: Tensile stress in the surface layer and compression in the substrate. The degradation of surface layer also leads to material embrittlement, i.e., the dramatic reduction of material toughness. A combination of growing tensile residual stress and decreasing toughness leads to a crack initiation at a certain level of degradation. Thus, no assistance of a defect is required for crack initiation in this case, though a presence of a defect in the surface layer may accelerate the initiation.

A characteristic feature of degradation-driven fracture initiation is closely spaced multiple cracking as illustrated in Figure 5.14. The photo in Figure 5.14a presents a cross section through polybutylene (PB) pipe wall (original PB pipe has a blue color) and a view of the inner surface of the pipe. A small part of the surface on the left from the dark grove was covered by fitting. It was protected from aggressive environments, and thus was preserved the original color. The grove is left from the edge of the fitting. On the right of the grove, one can see degraded inner surface, which was in contact with aggressive for PB fluid. A larger magnification of the small domain identified by box U is shown in Figure 5.14b. Here, a thin white layer with white spikes emanating from it in the vertical direction is the degraded PB. The white color of degraded resin results from the bleaching effect of aggressive fluid. The black spot in the center of the white layer in Figure 5.14b is a lost chunk of the very brittle degraded PB. Multiple closely spaced microcracks are initiated within the layer. As these microcracks grow, some of them grow faster

Figure 5.14 Multiple cracks emanating from the degraded surface layer in polybutylene pipe.

than the others grow and shield (terminate) the neighboring cracks. As this process progresses, one or a few cracks break through the pipe wall causing macroscopic failure of the pipe visible on the left side of Figure 5.14a. The densification of the heavily degraded PB is relatively small, about 1–1.5% of density increase. The degraded surface layer is also very thin, and it is about 50–75 μm. However, even such a small variation of density within a very thin surface layer turns to be sufficient to build up tensile stresses approaching the yield strength of original material (Choi et al., 2003). The rate of PB degradation process in potable water distribution services depends on a number of factors such as an inevitable variation of material composition on micron scale, fluid oxidative potential, local perturbation of the flow pattern, pressure, and temperature. Thus, the degradation-driven fracture initiation is at least as uncertain as that for the stress-driven fracture. As a result, a probabilistic method is the most adequate way to characterize and model brittle fracture initiation and growth [Chudnovsky and Kunin, 1987].

Mechanisms and Kinetics of Slow Crack Growth

Crack propagation in plastic components under fatigue and creep conditions usually appears as a narrow cut with small amplitude random deviations from a straight or slightly curved trajectory. A close observation of a crack reveals the presence of a process zone (PZ) (also called damage zone, plastic deformation zone, etc.) that precedes and surrounds the crack. Depending on the chemical composition and morphology of polymer, temperature, specimen geometry, and loading conditions, various types of micro-defects such as crazes, shear bands, microcracks, and micro-voids, may constitute PZ. In general, microdefects are strain localizations on microscale and, based on their orientation, one can distinguish brittle, such as crazes and microcracks, and ductile, like shear bands and micro-voids, types of damage. The microdefects are formed in response to stress concentration and shield the vicinity of the crack front from high elastic stresses by the increase of an effective (inelastic) material compliance. The crack growth is closely coupled with formation and evolution of the microdefects within PZ. A system of crack and PZ is referred to as Crack layer (CL). The cross section reveals an array of crazes well aligned with the crack faces. Apparently, it also affects the stress intensity factor (SIF) K. Figure 5.7 illustrates the effect of crazing (displacement discontinuities) on SIF. An individual craze is modeled as a displacement discontinuity across an elementary segment (area in 3D) with orientation n and opening b. In the specific example shown in Figure 5.7, all crazes are considered to be parallel to the crack and have an opening in the normal direction.

The illustration of CL propagation is shown in Figure 5.15 as a snapshot of CL in polystyrene and a sketch of AZ leading edge migration from one configuration to another traced from the video recording of fatigue CL

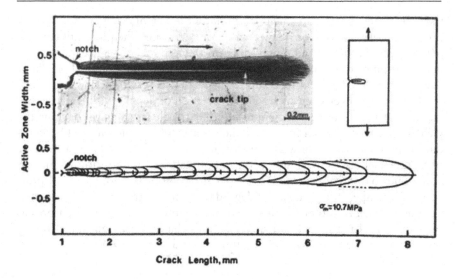

Figure 5.15 Schematics of CL evolution as a sequence of AZ configurations.

growth in single edge notched (SEN) specimen (shown in right upper corner). The small vertical marks along the horizontal axis indicate the crack tip positions corresponding to the sequence of the AZ leading edges. As it can be seen from Figure 5.15, the movement of AZ includes an isotropic expansion and distortion in addition to the ridged translation followed by the crack tip extension [Chudnovsky et al. (1988)]. The AZ evolution is inseparable from the crack growth. Another example of PZ constituted by a set of shear bands typical for ductile micro-mechanism of brittle fracture is shown in Figure 5.16. It illustrates a CL formed under fatigue in polycarbonate (PC), [Chudnovsky, A. (1984)] an amorphous thermoplastic. Figure 5.16 shows optical micrographs of the side view of PZ that surrounds and extends far ahead of the crack. Figure 5.16b displays the PZ cross section perpendicular to the crack face near crack tip at the location indicated by vertical arrow. The cross section reveals an array of intersecting shear bands (dark lines at about $\pm 45°$ to the vertical axis) and resulting in thinning of the specimen. The dark curve that outlines the contour of PZ in Figure 5.16a results from the light reflection effect from a curved side surface of the specimen resulting from thinning as depicted in Figure 5.16b. The aforementioned two examples illustrate CL with brittle (crazes) and ductile (shear bands) type of defects constituting PZ in plastics. The type of defects that constitute PZ depend on various factors such as chemical composition of material, temperature, specimen thickness, as well as morphological variations across the thickness, e.g., crystal size distribution in semicrystalline polymers. The effect of morphology is illustrated in

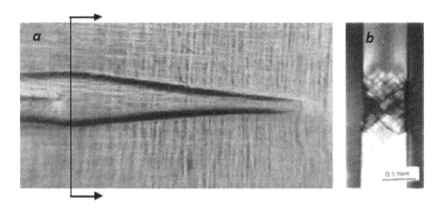

Figure 5.16 Side view of CL (a) and the cross section near the tip of AZ (b) in PC.

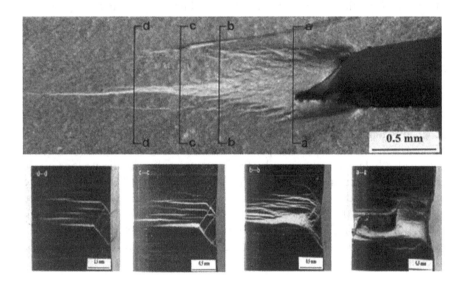

Figure 5.17 Optical micrographs of the side view and four cross sections of CL in PB.

Figure 5.17 by the micrographs of CL developed under fatigue loading at 110°C in PB, a semicrystalline thermoplastic. The specimen was cut from the extruded PB tubing, and as a result the inner surface layer (skin) has a different crystalline structure than that in the core and outer skin due to a different cooling rate inside (slow) and outside (fast). The morphological differences are manifested in different mechanical responses to stresses: ductile (shear bending) within the inner skin and brittle one (crazing) in the core and outer skin.

The cross section a–a displays a complete break and wide crack opening on the outer skin (left side of the lower set of micrographs) and a part of the core in contrast with intact inner skin and an adjacent part of the core with large deformation. Thus, the crack front extends farther on the outer side of PZ than on the inner one. Large white area on the optical micrograph represents a very intensive cavitation, crazing, and shear banding (white color results from light scatter). There are also a few visible discrete white lines. Higher resolution SEM micrographs reveal that these are the bundles of large crazes and/or shear bands. There is a lesser domain of large deformation and more strain localizations (white lines) in form of shear bands within the inner skin and crazes on the rest of the cross section b–b than that observed in a–a. As we move further toward the front of AZ, the number of discrete localizations (crazes and shear bands) decreases, and their thickness reduces. It means the damage density reduces toward periphery of AZ.

It should be noticed that no damage, i.e., no PZ has been observed in front of the crack in the same PB in the test conducted at room temperature in contrast with previously described AZ observed at 110°C. Thus, the mechanism of crack growth in this case dramatically changed with temperature. It creates an obstacle for the extrapolation of high-temperature accelerated test data to room temperature since the requirement of fracture mechanisms similarity is violated. The reproducibility of fracture mechanism is an important requirement for accelerated testing. In what follows, we report experimental examination of the similarity of the mechanism and kinetics of CL growth at various load levels and temperatures in a commercial polyethylene (PE). PE is a semicrystalline thermoplastic from polyolefin family. PB discussed before also belongs to that family. But let us focus for the discussions on PE in subsequent section since (a) it displays the simplest PZ in comparison to other plastics and the same time has main characteristic features observed in other plastics; (b) PE has the widest spectrum of engineering applications ranging from tubing for water distribution and high-pressure pipes for natural gas transmission lines to widely used packaging and various components of orthopedic prosthesis.

Thermodynamics and Phenomenological Modeling of CL

Studies of crack propagation and stability have developed in two main directions. One, related to materials science, concerns studies of the hierarchy of microdefects, their nucleation, interaction, and development in association with the propagation of a main crack. The microdefects are very different by nature depending on material structure and method of observation, specifically the level of magnification chosen. Using a progressively finer scale of observation, a hierarchy of defects can be visualized. For instance, the various elements of damage in the vicinity of a fatigue crack in AS1 301 stainless steel are shown in Figure 5.18. A zone of a large plastic deformation surrounding the crack tip appears under 25× (Figure 5.18a). The randomly oriented "turbulent" field of lines representing the localization of deformation can be distinctly observed under 500× (Figure 5.18b). Elements of discontinuity constituting an essential part of overall deformation can be identified under 20,000× (Figure 5.18c). Individual dislocations, the atomic structure, etc., can be observed under larger magnification. Which elements of this hierarchy of defects should be parametrized in order to be included into a quantitative model of a crack surrounded by damage? Apparently, this question should not be addressed by materials science only. Conventionally in continuum mechanics, a crack is considered as an ideal cut in an elastic, elastoplastic, or visco-elastoplastic medium. The concept of crack-cut with associated surface energy was the first and a very important step in studies of brittle failure [Griffith, A.A. (1921)]. It reflects some essential features of fracture processes and has served as a solid foundation for many engineering and scientific applications. Following this approach, microdefects surrounding the crack are modeled as a plastic zone in a very general sense. Such modeling describes the macroscopic deformation and stress state reasonably well for plastic metals. It does not, however, describe the [Griffith, A.A. (1921, 1925)] microdefect zone in polymers like PS, PE, and PB. It does not, however, describe the microstructure of plastic deformation within the "plastic zone" (see, e.g., Figure 5.18b and c). For brittle materials such as ceramics and rocks, the models of plasticity also seem inadequate. Recent achievements in materials science challenge the continuum approach to model the fracture processes. Obviously, the complexity of a hierarchy of interacting defects briefly mentioned earlier is the main obstacle. In order to simplify the picture, we may examine a crack surrounded by damage under relatively low magnification (Figures 5.19(a–c)) [Chudnovsky, A. and Bessendorf, M. (1983), Chudnovsky et al. (1983)]. These micrographs have been obtained for various materials: polystyrene (an amorphous polymer), polypropylene (a semicrystalline polymer), and stainless steel (a polycrystalline metal). For thermodynamic treatment, one can skip the microstructural details (addressed earlier through micromechanics). The purpose of this treatment is to derive a constitutive equation of CL propagation and separately bring in the micromechanics prediction of some of the parameters that appear in the equation.

The observations [Chudnovsky, A. and Bessendorf, M. (1983), Chudnovsky, A. and Moet, A. (1984), Chudnovsky et al. (1983)] indicate that under similar loading conditions, the global geometry and the evolution of an array of microdefects surrounding a main crack have many similar features for various materials in spite of all the differences in molecular structure and morphology. Fracture propagation is usually an irreversible process. Hence, the general framework of the thermodynamics of irreversible processes can be employed for modeling the phenomenon. A system of a crack and its surrounding damage is referred to as a Crack layer (CL). The theory of Crack layer propagation based on irreversible thermodynamics has been proposed

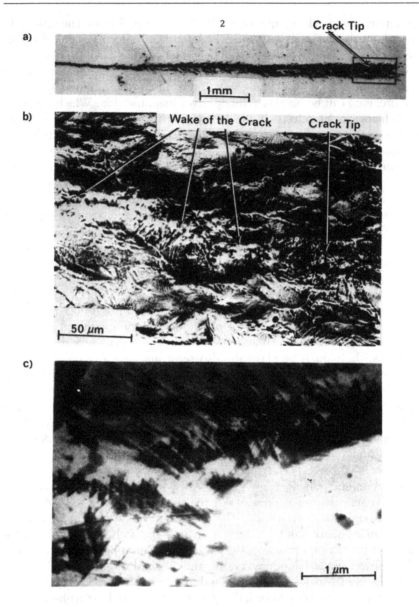

Figure 5.18 Morphology of the Crack layer. (a) General view of the fatigue Crack layer at low magnification. (b) The crack tip region. Extensive damage is seen around and in front of the crack tip. (c) SEM picture of an element of damage from the area in b taken at ~20,000 magnification.

in [Chudnovsky, A. Kiev 1976 (Abstract, in Russian); Chudnovsky et al. (1978); Khandogin, V. and Chudnovsky, A., Novosibirsk, 1978 (in Russian)] several pioneering papers. Supporting experimental evidence were reported

FATIGUE CRACK — POLYSTYRENE

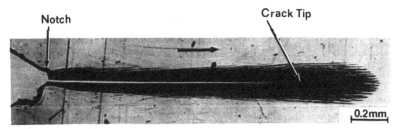

Fig. 2

FATIGUE CRACK — POLYPROPYLENE

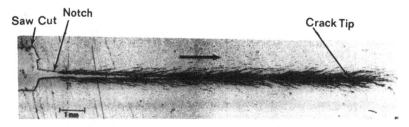

Fig. 3

FATIGUE CRACK — STAINLESS STEEL

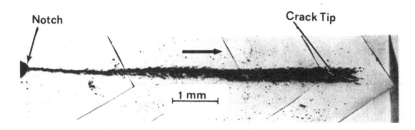

Figure 5.19 Micrograph of a crack propagating in (a) an amorphous polymer, polystyrene—damage zone surrounding the crack tip is called crazes, in
(b) semicrystalline polypropylene—damage zone surrounding the crack
tip is made of shear bands, and in (c) a polycrystalline metal—damage
zone surrounding the crack tip could be slip bands, dislocations, etc.

later [Chudnovsky, A. and Bessendorf, M., May 1983; Botsis et al., 4–10
December 1984; Chudnovsky et al., October 1983; Bakar et al. (1983), Botsis et al., 4–10 December 1984]. This work gives a comprehensive presenta-
tion of crack-layer (CL) as a model encompassing essential features of slow
fracture propagation in various materials.

Fracture processes consist of nucleation and growth of microdefects. For thermodynamic description of fracture, one needs to introduce a list of parameters of state by incorporating a damage parameter. To the authors' knowledge, L.M. Kachanov was the first one who did it explicitly [Kachanov, L.M. 1958 (in Russian).]. Since then, numerous papers suggesting various damage parameters and constitutive equations for them have been published. Specific interpretation of a damage is vitally important for establishing a correspondence between experimental studies and a damage model. It is not so important for general thermodynamic analysis, which is presented later in the chapter. However, it is always useful to have in mind a particular damage model. Therefore, in this section, we introduce a damage parameter (P) following a paper published in 1973 [Chudnovsky, A. 1973 (in Russian)]. Surfaces encompassing the discontinuities within an initially continuous solid are considered as the elements of damage (microcracks, crazes, shear bands, martensite transformation, etc.). A damage parameter P is defined as a pairing of scalar damage density ρ (i.e., an area of discontinuity surfaces per unit volume $[\rho] = m^2 / m^3$ and damage orientation parameter $O : P = \{\rho, O\}$). Thus, the following system of thermodynamic parameters of state is considered, $\{\tilde{\sigma}, T, P\}$.

Here, stress tensor $\tilde{\sigma}$ and the absolute temperature T are in the list of parameters of state of conventional elastic medium. This list is extended by adding the damage parameter P in order to describe the fracture processes. To deduce "thermodynamic causes" of damage, we derive the *entropy production* [de Groot S.R. and Mazur, 1969] associated with damage. The local energy balance can be written as:

$$\dot{u} = \tilde{\sigma} : \dot{\tilde{\varepsilon}} - \tilde{\nabla} \cdot \widetilde{j^Q} \tag{5.10}$$

Here, $\dot{u}$ stands for the rate of internal energy density, $\dot{\tilde{\varepsilon}}$ is a strain rate tensor, the product $\tilde{\sigma} : \dot{\tilde{\varepsilon}}$ represents the rate of work density, and $\tilde{\nabla} \cdot \widetilde{J^Q}$ gives the rate of internal energy density due to heat transfer ($\widetilde{J^Q}$ stands for heat flux). Considering small deformations, we decompose the total strain tensor into perfectly elastic $\tilde{\varepsilon}^{(e)}$ (thermodynamically reversible) and nonelastic $\tilde{\varepsilon}^{(i)}$ (irreversible) parts, i.e.,

$$\tilde{\varepsilon} = \tilde{\varepsilon}^{(e)} + \tilde{\varepsilon}^{(i)} \tag{5.11}$$

The work done on nonelastic deformation is spent partially on damage nucleation and growth and partially by being converted into heat. It can be expressed as follows: $\alpha\tilde{\sigma} : \dot{\tilde{\varepsilon}}^{(i)}$ is a part of the irreversible work associated with damage nucleation and growth and $(1-\alpha)\tilde{\sigma} : \dot{\tilde{\varepsilon}}^{(i)}$ is converted into heat, (α is a phenomenological coefficient). Let us consider the left part of Eq. (5.10). Conventionally, the internal energy density u consists of Helmholtz free energy density f and the entropic part Ts (following Eq. 5.12). Consequently,

$$\dot{u} = \dot{f} + T\dot{s} + s\dot{T} \tag{5.12}$$

According to the basic concept of irreversible thermodynamics, the time rate of the entropy density changes can be decomposed into two terms: $\dot{s} = \dot{s}_i + \dot{s}_e$, where $\dot{s}_i$ stands for the entropy production due to irreversible processes, and $\dot{s}_e$ is the entropy density rate due to a quasi-equilibrial exchanges with the surrounding by heat and other forms of energy. In the equilibrial thermodynamics, the entropy increment ΔS is defined as the ratio of heat/temperature. In non-equilibrial thermodynamics, the heat exchange with the surrounding is usually assumed to be equilibrial.

At this point, it is important to mention that the relationships between thermodynamic fluxes and forces were established by Ilya Prigogine. Prigogine [Prigogine, I., New York (1955/1961/1967); Prigogine, I. and Defay, R. (1950/1954); Prigogine, I. and Defay, R., London (1950/1954)] and others expressed the entropy production rate as a bilinear form of the thermodynamic fluxes and conjugated thermodynamic forces by the relation:

$$\Pr \dot{s}_i = \sum_k j_k^0 \cdot X_k^0 \tag{5.13}$$

where j_k^0 were called generalized fluxes and conjugated thermodynamic forces X_k^0 as generalized forces. Prigogine is best known for his definition of dissipative structures and their role in thermodynamic systems far from equilibrium, a discovery that won him the Nobel Prize in Chemistry in 1977. The authors of this book claim that the fracture propagation in an engineering structure is a dissipative process, and the engineering structure within which such a fracture processes are unfolding is just an example of such dissipative system and the same general thermodynamic treatment applies once the thermodynamic forces and fluxes are properly defined for the system in question [*Appendix IV*]. In fact, Lars Onsager's work predates Prigogine on this topic of reciprocal relations in irreversible processes and forms the foundation for the pioneering work on fracture processes that will be presented in this book for the first time [Onsager, L., i; Onsager, L., ii (1931)]. Onsager also received Nobel Prize in Chemistry in 1968 for the discovery of reciprocal relations (see *Appendix I* for historical remarks).

Going back to Eq. (5.12), we introduce the equilibrial entropy rate $\dot{s}_e$ in the following form:

$$\dot{s}_e = -\tilde{\nabla}\cdot\frac{1}{T}\widetilde{J^Q} + (1-\alpha)\tilde{\sigma}:\dot{\tilde{\varepsilon}}^{(i)} + \Delta S\cdot\dot{\rho} \tag{5.14}$$

Here, the first term represents the entropy rate due to the entropy flux $\tilde{J}^s = \frac{1}{T}\widetilde{J^Q}$, the second term reflects the entropy increases due to the heat generated by the irreversible work $\tilde{\sigma}:\dot{\tilde{\varepsilon}}^{(i)}$, and the third term reflects the rate of the entropy changes due to the localized transformation of the thermodynamic state such as cracking, crazing, and shear banding. $\Delta S\{\tilde{\sigma}, T, O\}$ stands for the difference between the entropies of damaged and undamaged matter. Since the stress tensor and the absolute temperature constitute the conventional part of the list of parameters of state, it is convenient to use Gibbs free energy density g. We define g as the difference between Helmholtz free energy density f and the density of the work done on elastic deformation $\tilde{\sigma}:\dot{\tilde{\varepsilon}}^{(e)}$. Consequently,

$$\dot{g} = \dot{f} - \tilde{\sigma}:\dot{\tilde{\varepsilon}}^{(e)} - \dot{\tilde{\sigma}}:\tilde{\varepsilon}^{(e)} \rightarrow \dot{g} + \dot{\tilde{\sigma}}:\tilde{\varepsilon}^{(e)} = \dot{f} - \tilde{\sigma}:\dot{\tilde{\varepsilon}}^{(e)} \tag{5.15}$$

Substituting Eqs. (5.11), (5.12) and (5.14) into (5.10) and solving the energy balance equation with respect to the energy production rate $\dot{s}_i$, and considering Eq. (5.15), we find:

$$T\dot{s}_i = \alpha\tilde{\sigma}:\dot{\tilde{\varepsilon}}^{(i)} - \dot{g} - \dot{\tilde{\sigma}}:\tilde{\varepsilon}^{(e)} - s\dot{T} - \Delta S\cdot\dot{\rho} - \frac{1}{T}\widetilde{J^Q}\tilde{\nabla}T \tag{5.16}$$

For further transformations, we decompose the rate $\dot{g}$ into two terms:

$$\dot{g}(\tilde{\sigma}, T, P) = \Delta g(\tilde{\sigma}, T, O)\dot{\rho} + \dot{v}(\tilde{\sigma}, T, P) \tag{5.17}$$

where Δg is the difference between the Gibb's free energy densities of damaged and undamaged matter, and Π is the elastic potential energy density. The assumption of local equilibrium yields the following conventional constitutive equations:

$$\left.\frac{\partial\Pi}{\partial T}\right|(\tilde{\sigma}, P) = -S \tag{5.18}$$

$$\frac{\partial \Pi}{\partial \tilde{\sigma}}\Big|\left(\tilde{T}, P\right) = -\tilde{\varepsilon}^{(e)} \tag{5.19}$$

We also make use of the conventional relationship between the densities of the enthalpy h, Gibbs free energy g, and entropy S in the following way:

$$\Delta g + T\Delta S = \Delta h\left(\tilde{\sigma}, T, O\right) \tag{5.20}$$

Then, substituting Eqs. (5.17–5.20) into Eq. (5.16) and having defined $\frac{\partial h}{\partial P}\widetilde{P^{def}} = \Delta h\dot{p}$, the following expression for the entropy production can be obtained:

$$T\dot{s}_i = v\tilde{\sigma} : \dot{\tilde{\varepsilon}}^{(i)} - \frac{\partial(h+v)}{\partial P}\dot{P} - \frac{1}{T}\widetilde{J^Q}\tilde{\nabla}T + \zeta \cdot A_f + \widetilde{J^m}\tilde{\nabla}C + \widetilde{J^m}\tilde{\nabla}P + \widetilde{J^m}\tilde{\nabla}C \tag{5.21}$$

The term α is a phenomenological coefficient representing a portion of stress–strain energy dissipated in internal friction. The third term in Eq. (5.21) describes the entropy production due to heat transfer which we will ignore in the rest of the development since it is only applicable to special case. Fourth term ζ represents chemical flux, and A_f represents reciprocal thermodynamic force, also known as chemical affinity (chemical affinity is the tendency of an atom or compound to combine by chemical reaction with atoms and compounds of unlike composition). This fourth term plays an important role in fracture phenomenon in a chemically aggressive environment, a field of mechano-chemistry. Fifth, sixth, and seventh terms play a role in filtration and diffusion processes where $\tilde{\nabla}C$ and $\tilde{\nabla}P$ represent concentration and pressure gradient (thermodynamic driving forces), respectively. Mass flux in filtration and diffusion is represented by J^m.

For simplicity, we will assume a situation where both these gradients $\tilde{\nabla}C$ and $\tilde{\nabla}P$ are zero for the time being. To concentrate attention on the damage process $\left(\dot{P}\right)$, we assume an isothermal condition and homogeneity of temperature field $\left(\tilde{\nabla}T = 0\right)$. Under these conditions, the entropy production 5.21 reduces to two terms, and the first term is directly associated with damage and therefore is nonzero only within a region where $\dot{P} \neq 0$. The second term represents the entropy production due to damage growth $\left(\dot{P}\right)$ directly. Therefore,

$$T\dot{s}_i = \alpha\tilde{\sigma} : \dot{\tilde{\varepsilon}}^{(i)} - \frac{\partial(h+\Pi)}{\partial P}\dot{P} \tag{5.22}$$

The rate of damage growth $\dot{P}$ may be considered as a thermodynamic flux. Then $-\dfrac{\partial(h+\Pi)}{\partial P}$ is the reciprocal force [Onsager, L., i; Onsager, L., ii (1931)].

If one visualizes the damage as a field of microcracks with microcrack density ρ, then the rate of the potential energy density with respect to ρ is always negative [Chudnovsky, A. 1973 (in Russian)]. It means that the contribution of the potential energy density change to the thermodynamic forces is always positive. It can be shown for the same conditions that the enthalpy increment Δh as well as the entropy increment ΔS due to cracking is always positive [Chudnovsky, A. 1973 (in Russian)]. Thus, the thermodynamic force $-\dfrac{\partial(h+\Pi)}{\partial P}$ reciprocal to the rate of damage $\dot{P}$ results from the competition between the driving $\left(-\dfrac{\partial\Pi}{\partial P}>0\right)$ and the resisting $\left(-\dfrac{\partial h}{\partial P}<0\right)$ parts. To emphasize the importance of the enthalpy increment for the thermodynamics of failure, we introduce the specific enthalpy of damage γ^* with respect to the unit of damage chosen $[\rho]=\dfrac{m^2}{m^3}$. The establishment of a constitutive equation for damage growth $\dot{P}$ would be the most desirable goal. This requires an effort to specify the damage parameter $\dot{P}$ and microscopical observation of damage evolution under various loading conditions. In addition, for crack

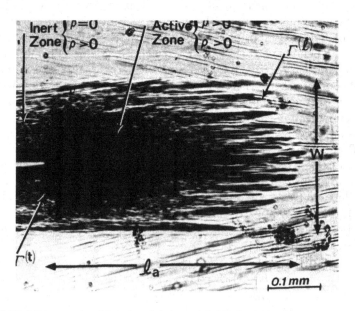

Figure 5.20 Micrograph of the tip of the crack in PS distinguishing active and inert zones.

propagation studies, integral characteristics of damage surrounding the crack rather than details of the damage distribution are important. Therefore, following the spirit of thermodynamics, we introduce average characteristics of the entire damage zone including the crack (*Crack layer*) and constitutive equations for the CL without reference to a constitutive law for local damage. Obviously, the constitutive equations for the Crack layer could be deduced from a constitutive law of local damage.

The concept of CL considers a system of a crack and the surrounding array of microdefects as one macroscopic entity. As an illustrative example, the trace of fatigue CL propagation in polystyrene (PS) is shown in Figure 5.20. The damaged zone expands in a self-similar fashion. The observation of damage (in the case of PS, the damage constitutes crazes) in the vicinity of the crack tip [Kramer, E.J., July 1984; Botsis et al., 4–10 December 1984] suggests that the damage distribution appears as a manifestation of actual stress field. Formally, a Crack layer is described as zone V within which the damage density ρ is positive. It is worth noting that a certain level of damage could preexist, independent of crack propagation. In that case, ρ_0 should be determined as a reference level. The CL is then defined as a zone within which the damage density is above the reference level. To describe CL propagation, we distinguish active and inert zones (see Figure 5.40) within the CL. In a zone adjacent to the crack tip, the damage density keeps growing under the influence of stress concentration. The zone, V_A, within which the damage density is above the reference level ρ_0 and the rate of damage density is positive, is called the active zone (see Figure 5.20). When a crack propagates through the active zone, the stresses are released, and consequently the process of damage growth practically stops. Thus, the inert zone appears as a trace of the active zone propagation. The inert zone V_I is complementary to the active zone V_A. The active zone boundary ∂V_A can be represented as consisting of the leading Γ^l and trailing Γ^t edges (see Figure 5.20). The trailing edge Γ^t is defined as the border between the active and inert zones. The leading edge Γ^l is the part of ∂V_A complementary to Γ^t. A characteristic width w_a and length l_a of the active zone are shown in Figure 5.20. For simplicity, we consider the case when both w_a and l_a are small in comparison with the main crack length ℓ.

The mechanism of Crack layer propagation can generally be described as follows. At first, damage nucleates and grows to a critical level at which the local instability conditions are met [Chudnovsky, A. 1973 (in Russian)]. Then a crack appears within the damage zone. The crack creates a stress concentration which intensifies the processes of damage growth. The damage density in front of the crack reaches a critical level which leads to crack extension and so on. The critical level of damage relates to the stresses through the conditions of instability [Chudnovsky, A. 1973 (in Russian)]. Since the stress distribution in the vicinity of a crack has an invariant shape with respect to crack length (the crack length just scales the stresses by a stress intensity factor K). The critical level of damage is maintained constant during the crack propagation. The CL propagation can be visualized as an active zone (damage distribution) movement. The latter can be decomposed into translation and rotation as a rigid body and an active zone deformation (i.e., damage dissemination). Considering homogeneous deformation only, one can express the rate of the damage parameter $\dot{P}$ resulting from the active zone movement without explicit time dependency in the following form:

$$\dot{P} = \dot{v}\,\delta^{tr}\,P + \dot{w}\,\delta^{rot}\,P + \dot{e}\,\delta^{exp}\,P + \dot{d}\,\delta^{dev}\,P \tag{5.23}$$

Here $\dot{v}$, $\dot{w}$, $\dot{e}$, and $\dot{d}$ stand for the rates of translation, rotation, isotropic expansion, and deviatoric deformation, correspondingly. δ^{tr}, δ^{rot}, δ^{exp}, and δ^{dev} are the operators of translational, rotational, expansional, and deviatoric transformations, respectively. Global entropy production due to the Crack layer propagation is the integral over the entire volume of a solid V:

$$s_i^{global} = \int_V \dot{s}_i\,dV \tag{5.24}$$

Since the entropy production due to damage growth is nonzero only within the active zone, the integral is reduced into the integral over V_a. Further substituting 5.23 into 5.21, and integrating we obtain,

$$Ts_i^{global} = \dot{D} + \tilde{v} \cdot \widetilde{X^{tr}} + \tilde{w} \cdot \widetilde{X^{rot}} + \dot{e} \cdot \widetilde{X^{exp}} + \tilde{d}\widetilde{X^{dev}} \tag{5.25}$$

Here, $\dot{D}$ is the part of irreversible work within the active zone spent on damage, i.e.,

$$\dot{D} = \alpha \int_{V_A} \tilde{\sigma} : \dot{\tilde{\varepsilon}}^{(i)} dV = \dot{W}^{(i)} + \dot{Q} \tag{5.26}$$

where $\dot{W}^{(i)} = \alpha \int_{V_A} \tilde{\sigma} : \dot{\tilde{\varepsilon}}^{(i)} dV$ is the rate of the total work dissipated within V_A and the rate of heat radiation $\dot{Q} = \int_{V_A} j^Q \cdot \tilde{n} dA$. The thermodynamic forces that are translational $\widetilde{X^{tr}}$, rotational $\widetilde{X^{rot}}$, expansional $\widetilde{X^{exp}}$, and deviatoric $\widetilde{X^{dev}}$ can be presented as follows:

$$\widetilde{X^{tr}} = \int_{V_A} \frac{\partial(h+\Pi)}{\partial P}\tilde{\delta}^{tr}P dV \tag{5.27}$$

$$\widetilde{X^{rot}} = \int_{V_A} \frac{\partial(h+\Pi)}{\partial P}\tilde{\delta}^{rot}P dV \tag{5.28}$$

$$\widetilde{X^{exp}} = \int_{V_A} \frac{\partial(h+\Pi)}{\partial P} \widetilde{\delta}^{exp} P dV \tag{5.29}$$

$$\widetilde{X^{dev}} = \int_{V_A} \frac{\partial(h+\Pi)}{\partial P} \widetilde{\delta}^{dev} P dV \tag{5.30}$$

In order to calculate the thermodynamic forces for the CL, we consider the enthalpic (resisting) and potential energy release (driving) parts separately. Driving parts require calculating the potential energy release rates. Equation (4.71) represents a transformation applied to the elastic potential energy density $\Pi(\tilde{\sigma}, T, P)$ using arbitrary operator "δ" and assuming homogeneous temperature field. Subsequent equations (4.72–4.75) finally arrive at J_1 which stands for conventionally used potential energy release rate with respect to crack extension perpendicular to the direction of applied stress [Eq. (4.75)]. In the same way, Eq. (4.79) expresses energy release rate with respect to CL rotation, L_m, and 4.82 expresses energy release rate with respect to CL expansion, M. As stated earlier, the M-integral in 4.82 possesses the path-invariancy for linear medium only. In a similar fashion, a second rank tensor N_{kl} can be introduced as the potential energy release due to deviatoric deformation. N_{kl} is particularly sensitive to deviations in the stress and strain fields near the crack tip. In contrast to j_k, L_m, and M, N_{kl} does not possess path independency. This means the last two integrals cannot produce any material parameter like J_{1c}.

Calculation of the resistance parts requires evaluating the enthalpic part of the thermodynamic forces and hence needs to specify the damage parameter. We make use of the damage parameter discussed earlier, $P=\{\rho,O\}$, where ρ is a scalar with dimension 1/(damage density), and O stands for an average damage orientation. If the orientation of an element of the damage (a plane microcrack, that is a plane surface element across which a discontinuity of originally continuous solid takes place it may be a microcrack or a craze). Thus, the damage density reflects an average distance between such elementary planes, for instance, within a RV. The orientation is described by a vector attached to the center of the element of the surface, then the translation and isotropic expansion transform the damage density P only and do not affect the orientation O. The rotation and the deviatoric deformation affect both the damage density and the orientation. We arrive at the change of the enthalpy h due to damage according to Eqs. (5.14), (5.17), and (5.20) and considering the relationship of γ^* and Δh:

$$\frac{\partial h}{\partial P}\delta P = \gamma^*\left(\widetilde{\sigma},T,O\right)\delta P + \rho\frac{\partial\gamma^*}{\partial O}\delta O \tag{5.31}$$

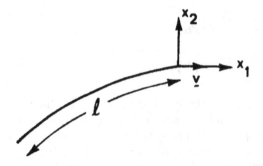

Figure 5.21 A smooth crack trajectory. Local Cartesian coordinate system has the origin at the crack tip "O", and the axis "OX$_1$" is directed along the tangent to the crack trajectory. ℓ represents the crack length.

Here, γ^* is the specific enthalpy of damage introduced earlier. General expression 5.31 can be used to derive the resistance parts of the thermodynamic forces. Substituting 152, the expression for the operator of translations into the first term of 148 and keeping in mind that a translation affects the damage density only, one can find:

$$-\int_{V_A} \frac{\partial h}{\partial P} \delta_k^{tr} P dV = \int_{V_A} \gamma^* \partial_k \rho dV = \gamma^* \int_{\partial V} \rho n_k d\Gamma = \gamma^* R_k \qquad (5.32)$$

where $R_k = \int_{\partial V} \rho n_k d\Gamma$ is defined as the translational resistance moment. Let us consider a smooth crack trajectory (Figure 5.21) and a local Cartesian coordinate system with the origin at the crack tip and the axis OX$_1$ directed along the tangent to the crack trajectory. Then, the crack speed vector v has only one nonzero component v$_1$ (v$_2$ = 0). Accordingly, only one component of the translational resistance moment R_1 is of interest. That is,

$$R_1 = -\int_{\partial V_A} \rho n_1 d\Gamma = -\int_{\Gamma(\ell)} \rho n_1 d\Gamma - \int_{\Gamma(\ell)} \rho n_1 d\Gamma = -\int_{\Gamma(\ell)} \rho_0 n_1^+ d\Gamma + \int_{\Gamma(\ell)} \rho n_1^- d\Gamma = \tilde{t} \int_{\Gamma} (\rho - \rho_0) dx_2 \ (5.33)$$

where n^+ stands for a unit vector normal to ∂V which has the positive or negative projection on the unit tangent $\tau((n,\tau) \lessgtr 0$, respectively (Figure 5.22).

The path of integration in 5.33 (the trailing edge) contains a singular point (the crack tip) where the damage density may be singular. Thus, we describe the damage density $\rho(x_2)$ along Γ^t as the sum of a singular $R_1^c \delta(x_2)$ and

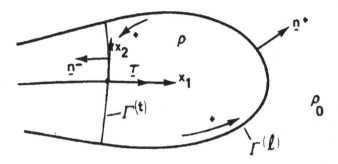

Figure 5.22 A schematic of the Crack layer front. Crack tip is at "O". The curve Γ^t marks the boundary between the wake zone and active zone ahead of the crack tip. Γ^ℓ represents the leading edge of the active zone. Local Cartesian coordinate system has the origin at the crack tip "O" and the axis "OX₁" is directed along the tangent to the crack trajectory. n^+ stands for a unit vector normal to Γ^ℓ and n^- stands for the unit vector normal to the trailing edge Γ^t. τ is the unit tangent to the crack trajectory.

regular $\rho^r(x_2)$ densities, i.e., $\rho(x_2) = R_1^c \delta(x_2) + \rho^r(x_2)$ respectively. R_1^c is defined as the core of damage. Substituting $\rho(x_2)$ into 5.33 and integrating, we find:

$$R_1 = R_1^c + R_1^r \qquad (5.33a)$$

In a similar fashion, substituting 5.31 and the expression for the operator of an infinitesimal isotropic expansion into the first term of 5.27 we obtain

$$\int_{V_A} \frac{\partial h}{\partial P} \delta_k^{exp} P \, dV = -\gamma^* \int_{V_A} x_k \partial_k \rho \, dV = \gamma^* R_0 \qquad (5.34)$$

Here, the expansion resistance moment R_0 is defined as:

$$R_0 = -\int_{V_A} x_k \partial_k \rho \, dV = -\int_{V_A} \lfloor \partial_k (x_k \rho) - \rho \rfloor dV = \int_{\partial V_A} \rho \, dV - \int_{\partial V_A} x_k \rho n_k d\Gamma \qquad (5.35)$$

In a case when the reference damage density $\rho_0 = 0$, the first term on the right-hand side of 5.35 can be neglected. Indeed,

$$\int_{\partial V_A} x_k \rho n_k d\Gamma = \int_{\Gamma^{(t)}} x_k \rho_0 n_k d\Gamma + \int_{\Gamma^{(t)}} x_k \rho n_k d\Gamma \sim 0 \qquad (5.35a)$$

Since $x_1 \sim 0$ and $n_2 \sim 0$ along the trailing edge Γ^t, therefore,

$$R_0 = \int_{\partial V_A} \rho dV = \langle \rho \rangle \cdot A \tag{5.36}$$

where $\langle \rho \rangle$ stands for an average over V_A damage density and A is the area of V_A. Substituting 5.31 and the expression for the operator of rotation into the first term of 5.28, we obtain

$$-\int \frac{\partial h}{\partial P} \delta_m^{rot} PdV = -\gamma^* R_m - \frac{\partial \gamma^*}{\partial O} R_m^0 \tag{5.37}$$

where R_m and R_m^0 stand for rotational resistance moments.

Assuming that the enthalpic barrier does not depend on the orientation of damage (a reasonable assumption for an isotropic medium), i.e., $\frac{\partial \gamma^*}{\partial O} = 0$, then 5.37 can be reduced into

$$-\int \frac{\partial h}{\partial P} \delta_m^{rot} PdV = -\gamma^* R_m = -\gamma^* R_m^{rot} \tag{5.38}$$

For deviatoric deformation of the active zone, a similar expression can be obtained:

$$-\int \frac{\partial h}{\partial P} \delta_{k\ell}^{dev} PdV = -\gamma^* R_{k\ell} \tag{5.39}$$

where $R_{k\ell} = (1 - \delta_{k\ell})^{1/2} \int_{V_A} (x_k \partial_\ell + x_k \partial_\ell) \rho dV$ (5.40)

and $\delta_{k\ell}$ stands for Kronecker Delta symbol. Finally, summarizing the results of this section, one can express the CL thermodynamic forces in an index form, which is convenient for practical application.

$$X_1^{tr} = \left(-J_1 + \gamma^* R_1 \right) \tag{5.41}$$

$$X_3^{rot} = \left(L_3 + \gamma^* R_m^{rot} \right) \tag{5.42}$$

$$X^{exp} = \left(M + \gamma^* R_0 \right) \tag{5.43}$$

$$X_k^{dev} = \left(N_k - \gamma^* R_{k\ell} \right) \tag{5.44}$$

It should be noted that the singular damage density $R_1^c \delta(x_2)$ contributes to translational resistance R_1 only, as it produces no effects for R_0, R_m^{rot}, or $R_{k\ell}$.

Rectilinear CL Propagation (Single Parameter Model)

To analyze the constitutive laws of Crack layer propagation, we start from the simplest case when all degrees of freedom are frozen except one. We assume that Crack layer propagation appears as a translation of the active zone along a rectilinear path with neither deformation nor rotation. Therefore, $\dot{w} = 0$, $\dot{e} = 0$, $\dot{d} = 0$, and $v_2 = 0$ (in terms of the coordinates of Figure 5.28). Since the crack trajectory is rectilinear, $v_1 = \dot{\ell}$ where ℓ is the crack length. Using 5.31, the global entropy production 5.25 can be rewritten as follows:

$$Ts_i^{global} = \dot{D} + \dot{\ell} \cdot \left(J_1 - \gamma^* R_1 \right) \tag{5.45}$$

According to the second law of thermodynamics, the entropy production (global as well as local) is nonnegative and equals zero for reversible processes. This does not mean that the processes causing negative entropy production $\dot{S}_i$ do not take place at all. Such processes are well known in chemical thermodynamics [Haase, 1963 (in German); Glansdorff, P. and Prigogine, I. (1971)]. The negative entropy-producing processes may occur if some other dissipative processes produce a sufficient amount of entropy to make the total entropy production nonnegative. In this case, the rates of entropy-consuming processes are controlled by other sources of entropy production. The constitutive law can therefore be obtained from the first principles. This is the case for the Crack layer propagation.

   In order to derive the law of rectilinear CL propagation, we need to analyze the stability of the CL. Assuming that there are no other sources of dissipation except of CL growth, i.e., $\dot{D} = 0$, the global entropy production takes the form

$$Ts_i^{global} = \dot{\ell} \cdot \left(J_1 - \gamma^* R_1 \right) \tag{5.46}$$

In classical thermodynamics, the instability condition is the prerogative of the second law. There are no disagreements about the criteria of instability of an equilibrial state. However, it is not so clear what instability criteria should be accepted for an irreversible process. In this study, we make use of the "universal criterion of evolution" which has been recently proposed and successfully applied to various irreversible processes [Glansdorff, P. and Prigogine, I. (1971)]. Since the entropy production is a bilinear form of thermodynamic fluxes j_k and reciprocal forces X_k:

$$\dot{S}_i = \sum_k j_k \cdot X_k \tag{5.47}$$

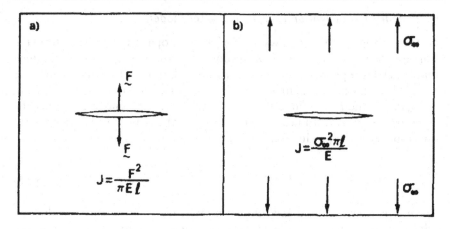

Figure 5.23 Center crack in an infinite plate (a) loaded at the crack face with force F and (b) stress σ^{∞} applied at infinity.

The time rate of the entropy production can be naturally decomposed into two terms:

$$\frac{d\dot{S}_i}{dt} = \frac{d_j \dot{S}_i}{dt} + \frac{d_x \dot{S}_i}{dt} \tag{5.48}$$

where

$$\frac{d_j \dot{S}_i}{dt} = \sum_k X_k \cdot \frac{dj_k}{dt} \tag{5.49}$$

and

$$\frac{d_x \dot{S}_i}{dt} = \sum_k j_k \cdot \frac{dX_k}{dt} \tag{5.50}$$

The criterion of evolution [Glansdorff, P. and Prigogine, I. (1971)] states that the following inequality

$$\frac{d_x \dot{S}_i}{dt} \leq 0 \tag{5.51}$$

always holds true for stable processes. The equality is met for either a stationary process or a critical situation when a sudden (uncontrolled) transition becomes possible. Analysis of the second variation $\delta^2 \dot{S}_i$ is necessary to distinguish these cases. Let us apply this criterion to the entropy production due to Crack layer extension (5.46):

$$\frac{d_x \dot{S}_i^{global}}{dt} = \ell^2 \cdot \frac{d}{d\ell}(J_1 - \gamma^* R_1) = \ell^2 \frac{dJ_1}{d\ell} \leq 0 \tag{5.52}$$

Indeed, the crack length ℓ is the only variable, since $\bar{\sigma}$ is constant. R_1 is constant according to our assumption (there is no deformation of the active zone). From 5.52 one can conclude that $\dfrac{dJ}{d\ell}$ should be < 0 for a stable crack growth. Two types of crack configuration can be distinguished from the stability viewpoint: $\left(\dfrac{dJ}{d\ell} < 0\right)$ stable and unstable $\left(\dfrac{dJ}{d\ell} > 0\right)$. Illustrative examples of these two distinct types are shown in Figure 5.23. CL exhibits different behaviors for stable and unstable configurations. For the configuration shown in Figure 5.23a, the energy release rate J is given by $\dfrac{F^2}{\bar{E}\pi\ell}$. Consequently $\dfrac{dJ}{d\ell}$ is always negative, i.e., the configuration is always stable.

Then, the requirements of the second law (5.46) are only "controller" of CL propagation. For $J_1 \geq \gamma^* R_1$, the requirement (5.46) is met, and a Crack layer is "allowed" to grow with undefined speed $\dot{\ell} > 0$. To specify the speed, we consider stationary CL growth. The condition of stationarity, i.e., the equality in (5.51) yields

$$J_1 = \gamma^* R_1 \tag{5.53}$$

Then, the applied load F controls the crack propagation velocity. Indeed, from (5.52) and constancy of $\gamma^* R_1$, one can deduce:

$$\frac{d}{dt}\left(J_1 = \gamma^* R_1\right) \rightarrow \frac{\partial J}{\partial F}\dot{F} + \frac{\partial J}{\partial \ell}\dot{\ell} = 0 \tag{5.54}$$

Substituting $J = \dfrac{F^2}{\bar{E}\pi\ell}$ into Eq. (5.53), we obtain

$$\dot{\ell} = \frac{2\ell}{F}\dot{F} \tag{5.55}$$

The condition (5.53) is obviously in agreement with the principle of minimum entropy production [Knowles, J.K. and Sternberg, E. (1972)]. If $J < \gamma^* R_1$, the CL growth (i.e., $\dot{\ell} > 0$) consumes entropy (negative entropy production). According to the second law, it is possible if the dissipation $\dot{D}$ is sufficient to compensate the negative term $\dot{\ell}\cdot\left(J_1 - \gamma^* R_1\right)$. Applying the principle of minimum entropy production (the minimal value of s_i^{global} is zero), we find:

$$\dot{D} + \dot{\ell}\cdot\left(J_1 - \gamma^* R_1\right) = 0 \tag{5.56}$$

From which we arrive at the following equation of SCG:

$$\dot{\ell} = \frac{\dot{D}}{\left(\gamma^* R_1 - J_1\right)} \tag{5.57}$$

Summarizing the cases discussed, one can write a general expression for a stationary CL propagation in a stable configuration:

$$
\dot{\ell} = \begin{cases} \dfrac{\dot{D}}{\left(\gamma^* R_1 - J_1\right)} \text{ If } J_1 < \gamma^* R_1 \\[3mm] -\dfrac{\partial F}{\partial J} \dot{F} \text{ If } J_1 = \gamma^* R_1 \\ \quad \dfrac{\partial J}{\partial \ell} \end{cases} \tag{5.58}
$$

The upper solution in 5.58 is for dissipation-controlled process. The lower solution is for equilibrial crack growth.

CL Propagation in an Unstable Configuration

For the configuration shown in Figure 5.23b, the energy release rate is $J = \dfrac{\sigma_\infty^2 \pi}{E} \ell$. Consequently, $\dfrac{\partial J}{\partial \ell} = \dfrac{\sigma_\infty^2 \pi}{E}$ is always positive, i.e., the configuration is always unstable. The limitation introduced by the second law prevents the crack from a rapid avalanche-like propagation, i.e., the crack cannot lead to a catastrophic failure of an engineering structure if $Ts_i^{global} = \dot{\ell} \cdot \left(J_1 - \gamma^* R_1\right) < 0$ which yields for $\left(\dot{\ell} > 0\right)$ the following requirements for stable CL growth:

$$
J_1 - \gamma^* R_1 < 0 \tag{5.59}
$$

Therefore, a slow CL propagation, which is compatible with 5.59 can occur due to the dissipative term $\dot{D}$ only. Therefore, if one accepts the principle of minimum entropy production, then Eq. (5.56) and (5.57) holds true for an unstable configuration as well as for a stable one.

When J_1 approaches $\gamma^* R_1$, the requirement of the second law is met. Then for an unstable configuration, the crack propagates avalanche-like manner if $J_1 = \gamma^* R_1$. Summarizing the results, one can write:

$$
\dot{\ell} = \begin{cases} \dfrac{\dot{D}}{\left(\gamma^* R_1 - J_1\right)} \text{ If } J_1 < \gamma^* R_1 \\[3mm] \text{undefined If } J_1 = \gamma^* R_1 \end{cases} \tag{5.60}
$$

The upper solution in 5.60 is for dissipation-controlled process. The lower solution is undefined due to transition to the dynamic process. Although both characteristics $\dot{D}$ and $\gamma^* R_1$ can be studied using either stable or unstable configuration, the first is more convenient for analysis of the dissipation $\dot{D}$ while

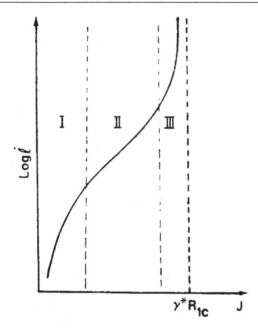

Figure 5.24 Three stages of slow crack propagation is schematically presented in a plot of ℓ versus J_1. The first stage corresponds to crack propagation through previously damaged material. The second, intermediate, stage is characterized by monotonic growth of damage accompanied by crack propagation. It is reflected in a monotonic increase of ℓ, with increasing J_1. The third stage is the subcritical stage of CL propagation.

unstable configuration is preferable for evaluation of $\gamma^* R_1$. The latter can be obtained from the conventional fracture toughness test and additional microscopic studies of damage distribution along Γ'. The rate of dissipation within the active zone $\dot{D}$ can be experimentally measured as the difference between the rate of total dissipated work $W^{(i)}$ and the heat $\dot{Q}$ radiated (see 5.26).

Slow CL propagation in unstable configuration (Eq. 5.60) is considered as an illustrative example. Since we do not have experimental data on heat radiation, we assume that the rate of dissipation $\dot{D}$ is proportional to the dissipated work W^i. The coefficient of proportionality apparently depends on mechanisms of dissipation. One would expect the coefficient to be dependent upon strain rate, temperature, and a characteristic time of the process. An evaluation of the irreversible work W^i due to an array of crazes constituting the CL active zone has been done in Botsis et al. (4–10 December 1984). The evaluation is based on a new CL stress analysis and an experimental CL characterization [Botsis et al., 4–10 December 1984; Botsis et al. (1987)].

These results suggest that W^i is proportional to the product $J\langle d\rangle$, where $\langle d\rangle$ stands for a characteristic size of V_A. Thus, the rate of dissipation $\dot{D}$ can be written as follows:

$$\dot{D} = \beta\langle d\rangle J \qquad (5.61)$$

Here, β is a phenomenological coefficient with dimension $[\beta]sec^{-1}$. At fixed temperature, β depends on the strain rate. The latter can be expressed in terms of the rate of applied load and a dimensionless crack propagation rate $\dot{\ell}/\langle d\rangle$ [Botsis et al., 4–10 December 1984]. Substituting (5.61) into (5.60), we obtain

$$\dot{\ell} = \begin{cases} \dfrac{\beta\langle d\rangle J}{\left(\gamma^* R_1 - J_1\right)} \text{ If } J_1 < \gamma^* R_1 \\[2mm] \text{undefined If } J_1 = \gamma^* R_1 \end{cases} \qquad (5.62)$$

The relationship 5.62 is schematically represented in Figure 5.23. One can distinguish three stages of slow crack propagation in unstable configuration (Figure 5.24). The first stage corresponds to crack propagation through previously damaged material. A new damage is not being developed at this stage. It means that the resistance moment R_1 mainly consists of the core of damage R_1^c which is small in comparison with a developed CL resistance R_1. Therefore, the rate of crack acceleration with respect to the energy release rate J_1 is relatively high. This is indicated by the slope of the stage I portion of the curve in Figure 5.24.

The second, intermediate, stage is characterized by monotonic growth of damage accompanied by crack propagation. It is reflected in a monotonic (approximately linear) increase of the translational resistance moment R_1 and the characteristic length with increasing of J_1. In this case, the translational driving force $\tilde{X}^{tr}$ is maintained approximately constant and the rate of crack propagation (5.60) can be approximated by a power type equation $\dot{\ell} \sim J^2$ $\left(i.e., K^4\right)$. The third, subcritical stage of CL propagation is characterized by deceleration of the resistance moment R_1 with respect to J. This results in the translational force $X_1^{tr} = \left(\gamma^* R_1 - J_1\right)$ approaching zero, which corresponds to the critical state (the asymptote in Figure 5.10). When $\left(\gamma^* R_1 - J_1\right) = 0$, the requirement of the second law (5.36) is met, and instability becomes permissible. Thus, at the end of the third stage, a SCG transforms into uncontrolled (avalanche-like) crack propagation. Obviously, the value of J_1 at which the translational force $\gamma^* R_1 - J_1$ equals zero corresponds to the critical energy release rate J_c (or G_c) in conventional Fracture mechanics, i.e.,

$$J_{1c} = \gamma^* R_{1c} \qquad (5.63)$$

is a parameter which can be experimentally evaluated using Fracture mechanics. γ^* and R_{1c} can be measured by materials science methods. Thus, 5.63 suggests a link between the two approaches. It will be further discussed through actual examples. The simple model described earlier generally predicts the shape of ℓ versus J_1 curve. However, it does not describe crack deceleration phenomena, history dependency of J_1, etc.

Let us consider a two-parameter model of rectilinear CL Propagation. Limitations of the previous model appear due to the employment of a single parameter only (the crack length, ℓ). A natural way to overcome the limitations is to use an additional degree of freedom offered by the CL model. Following this idea, we consider Crack layer propagation by translation along the rectilinear path and isotropic expansion of the active zone. Similar to the previous case, one can write: $v_1 = \ell$; $v_2 - 0$ (rectilinearity); $\dot{w} = 0$ (no rotation),

$$\dot{e} = \frac{1}{2}\left(\frac{\dot{w}_a}{w_a} + \frac{\dot{\ell}_a}{\ell_a}\right) \text{ (isotropic expansion from the crack tip as an origin)}, \ \dot{d} = 0$$

(no deviatoric deformation).

The global entropy production now can be rewritten as follows:

$$Ts_i^{global} = \dot{D} + \dot{\ell} \cdot \left(J_1 - \gamma^* R_1\right) + \dot{e}\left(M - \gamma^* R_0\right) \tag{5.64}$$

The criterion of stability (5.51) for the Crack layer only as we did in the previous example can be written as follows:

$$T\frac{d_x S_i^{global}}{dt} = \ell^2 \cdot \frac{\partial}{\partial \ell}\left(J_1 - \gamma^* R_1\right) + \dot{\ell}\dot{e}\left[\frac{\partial}{\partial e}\left(J_1 - \gamma^* R_1\right) + \frac{\partial}{\partial \ell}\left(M - \gamma^* R_0\right)\right]$$
$$+ \left(\dot{e}\right)^2 \frac{\partial}{\partial e}\left(M - \gamma^* R_0\right) \le 0 \tag{5.65}$$

Since $\dot{\ell}$ and $\dot{e}$ are independent variables, the aforementioned expression yields:

$$\frac{\partial}{\partial \ell}\left(J_1 - \gamma^* R_1\right) < 0 \tag{5.66}$$

$$\frac{\partial}{\partial e}\left(M - \gamma^* R_0\right) \le 0 \tag{5.67}$$

$$\begin{bmatrix} \frac{\partial}{\partial \ell}\left(J_1 - \gamma^* R_1\right) & \frac{1}{2}\left[\frac{\partial}{\partial e}\left(J_1 - \gamma^* R_1\right) + \frac{\partial}{\partial \ell}\left(M - \gamma^* R_0\right)\right] \\ \frac{1}{2}\left[\frac{\partial}{\partial e}\left(J_1 - \gamma^* R_1\right) + \frac{\partial}{\partial \ell}\left(M - \gamma^* R_0\right)\right] & \frac{\partial}{\partial e}\left(M - \gamma^* R_0\right) \end{bmatrix} < 0 \tag{5.68}$$

Using the same argument as in the single parameter model, one can conclude that both the expansional $\left(M - \gamma^* R_0\right)$ and translational $\left(J_1 - \gamma^* R_1\right)$ forces are always nonpositive for a stable Crack layer propagation. Therefore, two types of Crack layer configurations (stable and unstable) can be distinguished. Incorporating the principle of minimum entropy production, one can analyze stationary Crack layer propagation following the formalism used in the single-parameter model.

The stationary Crack layer propagation controlled by dissipation is described by the same Eq. (5.56) for both stable and unstable configurations (compare (5.57) and (5.59)). For this reason, we consider next the dissipation-controlled propagation only. The principle of minimum entropy production for slow CL propagation can be expressed in the form:

$$T s_i^{global} = \dot{D} - \dot{\ell} \cdot \left(\gamma^* R_1 - J_1 \right) + \dot{e} \left(\gamma^* R_0 - M \right) = 0 \qquad (5.69)$$

It implies that the dissipation $\dot{D}$ is distributed between two entropy sinks associated with two independent degrees of freedom: $\dot{\ell}$ and $\dot{e}$. As in the previous case, we assumed the rate of dissipation $\dot{D}$ to be expressed by 5.61. In this case, two parameters β_1 and β_2 substitute β to describe the distribution of the dissipation between two degrees of freedom. Then, 5.69 yields

$$\dot{\ell} = \frac{\beta_1 \langle d \rangle J_1}{\left(\gamma^* R_1 - J_1 \right)} \qquad (5.70)$$

$$\dot{e} = \frac{\beta_2 \langle d \rangle J_1}{\left(\gamma^* R_0 - M \right)} \qquad (5.71)$$

Since the isotropic expansion is the only source for the resistance moment changes (it affects only the regular part R_1^r), one can write

$$\dot{R}_1^r = R_1^r (0) \dot{e} \qquad (5.72)$$

Accordingly, Eq. (5.71) can be converted into an equation for R_1^r evolution using 5.72:

$$\dot{R}_1^r = \frac{\beta_2 R_1^r (0) \langle d \rangle J_1}{\left(\gamma^* R_0 - M \right)} \qquad (5.73)$$

To analyze the changes of R_1^r with crack length, one can calculate $\dfrac{dR_1^r}{d\ell}$ by taking the ratio of (5.73) and (5.70):

$$\frac{dR_1^r}{d\ell} = k R_1^r (0) \frac{\gamma^* R_1 - J_1}{\gamma^* R_0 - M} \qquad (5.74)$$

where $k = \dfrac{\beta_2}{\beta_1}$ is a phenomenological coefficient. It can be shown that J_1 approaches $\gamma^* R_1$ faster than M approaches $\gamma^* R_0$ for unstable configuration. Therefore, $\dfrac{dR_1^r}{d\ell}$ vanishes when $J_1 = \gamma^* R_1$. Equation (5.74) suggests a law of

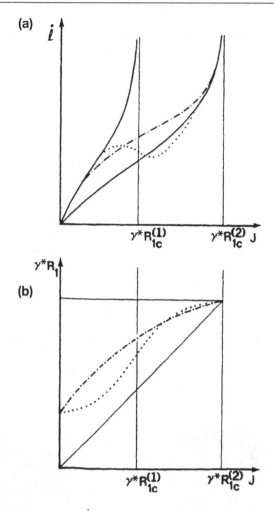

Figure 5.25 a A schematic of $\dot{\ell}$ versus J for two different CL growths in same material under two different loading conditions leading to different values of R_1^c based on the extent of damage. (b) A plot of $\gamma^* R_1$ versus J for the same CL growth showing the transition.

R_1^r evolution. The total translational resistance moment in Eq. (5.33a) displays a behavior similar to R_1^r due to the constancy of R_1^c. Since $\gamma^* R_1$ is not a constant in the two-parameter model, the crack growth rate versus J_1 essentially differs from that of the single parameter model. The solid lines (1) and (2) in Figure 5.25a represent the crack growth rate ($\dot{\ell}$ versus J_1) according to the single-parameter model with two different values of R_1, i.e., $R_1^{(1)} < R_1^{(2)}$, respectively. Evolution of R_1 (Figure 5.25b) yields a transition from the curve

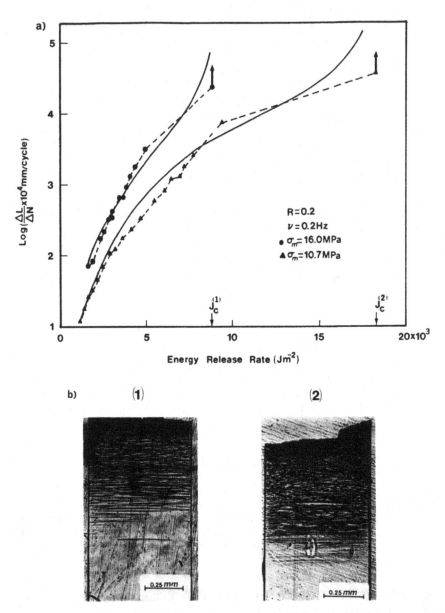

Figure 5.26 Plot of (a) crack growth rates ($\dfrac{\Delta l}{\Delta N}$ *in* log *scale*) versus energy release rate J and (b) transverse sections along the trailing edge at critical CL configuration $\left(J_c^1 / J_c^2 = R_c^1 / R_c^2 \right)$.

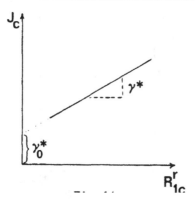

Figure 5.27 A schematic plot of J_{1c} as a function of R_1^r where γ^* represents the slope and γ_0^* represents the intercept.

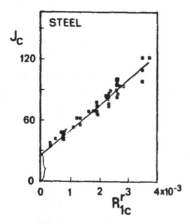

Figure 5.28 A plot of J_{1c} versus R_1^r in steel.

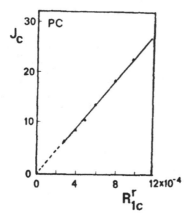

Figure 5.29 A plot of J_{1c} versus R_1^r in polycarbonate.

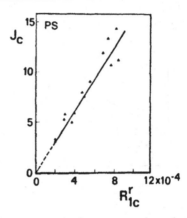

Figure 5.30 A plot of J_{1c} vs R_1^r in polystyrene.

(1) to the curve (2) (Figure 5.25a). Various ways of the transition depending on R_1 evolution are shown by dotted lines for monotonic transition and dashed line for a transition with crack deceleration (Figure 5.25). Applications of the two-parameter model to slow CL propagation in unstable configuration of polystyrene sheet under fatigue loading were achieved [Botsis et al., 1987/4]. Figure 5.26 taken from Botsis et al. (1987/4) shows a reasonable agreement between the theory and the experimental data for more than four order of magnitudes in crack growth rate. The two fatigue tests presented in Figure 5.26 were identical except of the values of σ_{mean}. The higher stress yields faster crack propagation and shorter fatigue time which was expected. Most important is the fact that the critical value of J_1 (at which avalanche-like crack propagation was observed) is much *smaller* for *higher stress*. The micrographs of the cross-section along the trailing edges of the two active zones shown in Figure 5.26b explain this phenomenon. The lower stress produces more dense damage, i.e., larger translational resistance moment, which consequently leads to larger J_{1c}.

Material toughness characterization in the two-parameter model also incorporates the necessary condition of CL instability already presented in 5.66, where R_1 is defined in 5.33a. When the sufficient condition of instability $\left(\dfrac{dJ}{d\ell} > 0 \right)$ is met, as it happens for various loading conditions, the condition 5.66 expresses the only requirement for the critical state. Following the conventional symbolism, we introduce J_{1c} as a critical value of J_1. Then using 5.66 and 5.33a, one can write,

$$J_{1c} = \gamma_0^* + \gamma^* R_1^r (t) \tag{5.75}$$

$\gamma_0^* = \gamma^* R_{1c}$ is a Griffith's type energy associated with either the crack surfaces or, in a more general sense, a near surface layer of intensive damage (a core of damage). During Crack layer growth, γ_0^* presumably remains constant. The second term describes the loading history dependency of J_{1c} and is associated with a damage dissemination accompanying the crack growth. Equation (5.75) presents J_{1c} as a linear function of R_1^r (Figure 5.27). The slope and the intercept in Figure 5.27 give rise to γ^* and γ_0^*, respectively. As shown in Figures 5.28, 5.29 and 5.30, the linear J_{1c}–R_1^r relationship has been demonstrated by the results on stainless steel [Chudnovsky, A. and Bessendorf, M., May 1983], polycarbonate, and polystyrene, respectively [Bakar et al., May 1983]. As it was suggested [Chudnovsky, A. 1973 (in Russian)], γ^* appears to be a constant quantity of the same order of magnitude as the latent energy of a phase transition for the material considered [Chudnovsky, A. and Bessendorf, M., May 1983; Bakar et al., May 1983].

In conclusion, the model described before suggests three independent parameters γ_0^*, γ^*, and R_{1c}^r to characterize material toughness. Two of them γ_0^* and γ^* are material constants, reflecting the mode of damage (microstructural features). The third is a history-dependent parameter. Therefore, a complete toughness characterization requires the establishment of the *constitutive equation* for R_1^r. The model described is in a good agreement with available experimental data. At the same time, the limitations of the model are obvious. It does not deal with the *crack trajectory* (we assumed a rectilinear path), it does not describe the active zone shape changes, observed recently [Chudnovsky et al., October 1983], etc. Therefore, the necessity of employing the rest of the Crack layer degrees of freedom is clear.

Crack Trajectory Considerations

There are two major issues in modeling the slow crack growth. The first one is related to the crack trajectory. The question is, how the crack trajectory is formed and/or what the criterion is for selecting local crack growth direction. The second issue is related to the crack growth rate along the selected trajectory. Most experiments in SCG studies have been conducted in such a way that cracks are propagated along a rectilinear path due to symmetry of the specimens and loading. This way one can eliminate the first issue and focus on the crack growth rate. Based on a number of studies, a rectilinear crack growth pattern has been established. Figure 4.11 in Chapter IV illustrates the complexity of crack growth process by emphasizing the existence of various stages of the process that may obey different "laws". Most of the empirical crack growth models steam from the Paris–Erdogan equation and relate the crack growth rate to the SIF or an effective SIF or most appropriately to energy release rate. Even the CL treatment for PS in earlier section is based on rectilinear crack growth which was forced by employing a V-shaped notch. The presence of a preexisting defect on the crack path may influence rectilinear crack growth. Chudnovsky et al. examined the case of crack trajectory experimentally by drilling a circular hole slightly below the rectilinear path leading to a crack–hole interaction problem in the same material PS. In addition to the load level reflected by SIF, there are three geometrical parameters of the crack–hole interaction problem: x-coordinate of the center of the hole, the distance from the reference line to the hole (in y-direction), and the hole diameter. Varying these parameters, they produced various degrees of perturbation of the stress field that crack propagates through impacting both the crack trajectory and the crack growth rate along that trajectory [Chudnovsky, A. (2014)]. Using computational techniques, they estimated the energy release rates for variable crack tip position along each of the observed trajectories. Comparison of the energy release rate for the crack with and without hole shows a significant difference when the crack tip is located close to the hole. Moreover, the trajectory consistently deviates toward the hole on the left of the deflection point and away from the hole on the right of the deflection point. This work also suggests that crazing (damage zone) has a strong effect on CL growth rate. At the same time, hole also initiates crazing at the boundary as the CL passes by. Thus, it is not only the CL that experiences the effect of the hole—the hole also, in turn, is affected by the CL. This type of situation is bound to occur in a real engineering structure when a crack is stopped at a hole that blunts the propagating crack. One can utilize this approach to stop a crack at early stage preventing catastrophic failure.

From examination of R-curve behavior discussed in the Fracture Mechanics section presented earlier, the variation of toughness parameter (K_{1c}) with crack size is expected since the resistance to crack growth depends on the energy dissipation within process zone, i.e., on the magnitude of the damage and the process zone dimensions. Moreover, the damage density [ρ] and active zone size depend on the fracture process and may vary depending on the loading history. This is clearly established in previous section using a crack in PS as an example. Thus, there is no basis to believe that the R-curve is unique for the given material, temperature, and specimen thickness, as suggested in ASTM E 561–98 Standard [Chudnovsky, A. (2014)]. However, the R-curve was not widely used in applications until the early 1970s. The main reasons appear to be that the R-curve depends on the geometry of the specimen and that the crack-driving force may be difficult to calculate [Erdogan, E. (2000)]. So far, we had been applying CL theory to a specific type of plastic, i.e., PS where damage zone constitutes crazes which are similar to microcracks. Specifically for this situation, the field of micromechanics had been introduced. However, some engineering material undergoes cold drawing as the damage process and different treatment are essential for such materials.

Crack Propagation Preceded by Cold Drawn Fibers

Engineering plastics like polyethylene (PE) undergo yield and cold drawing (also called necking) on the application of load instead of microcracking or crazing (as discussed earlier for amorphous polymers like PS). This cold drawing process occurs at the tip of the crack forming the damage zone. Any of the treatments of elastoplastic Fracture mechanics discussed earlier are useless for this type of damage. CL-driving forces depend on remote load, crack length, and AZ dimensions as well as the specimen geometry. Thus, one needs to monitor crack and AZ lengths in real time in order to formulate CL kinetic equations. Unfortunately, it is technically challenging to monitor CL dimensions in SCG process in PE partially due to the overall large deformation and existence of the side membranes that obstruct in-situ observation of AZ. A specimen geometry for which SIF does not depend on crack length allows one to simplify the observations and eliminate the complexity of the test data analysis resulting from variable SIF. Thus, the most convenient for studies of crack growth mechanism and kinetics are the specimens where SIF K_I and therefore the energy release rate (ERR) G_I depend only on the applied load and specimen geometry and do not depend on growing crack length. It stimulated a search for alternative specimen geometry. The search was conducted by means of finite elements method (FEM). It led to a stiff constant K (SCK) specimen [Chudnovsky et al. (2012)]. SCK specimen was chosen such that SIF due to remote load is independent of the crack length in the chosen range of crack size and within a defined range of SIF, simply related to the applied load in such geometry. Testing was done on specimens prepared from a large-diameter PE pipe. SCG tests were conducted under creep at various loads and temperatures.

SCG in PE

Under selected normalized stress levels and an elevated temperature, cracks grow faster and shorten PE lifetime in brittle fracture. However, at certain combination of load and temperature, a transition from continuous SCG to discontinuous, stepwise SCG has been reported [Chudnovsky et al. (2012), Zhou et al. (2010)]. The change in SCG mechanism is related to PE's ability to form a stable process zone (PZ) consisting of cold-drawn PE microfibers and micro-membranes in front of the crack. A high hydrostatic tension component of the crack tip stress field and PE propensity to cold drawing play a major role in PZ formation and growth. There are three main stages of brittle fracture: (a) crack initiation, (b) SCG, and (c) crack instability and transition to either rapid crack propagation (RCP) or ductile rupture. Note that the third stage occurs relatively fast, and the lifetime of the PE component mainly consists of time prior to crack initiation and the duration of stable SCG.

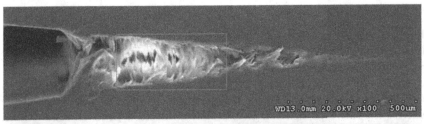

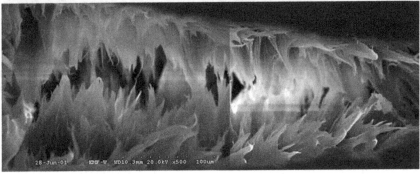

Figure 5.31 SEM micrographs of Crack layer in polyethylene: General view (×100) and part of wake zone (lower micrograph at ×500).

Crack initiation is the least studied stage in fracture (for more details, see paragraphs on *Fracture Initiation*) process. Just as in the case of PS, one needs to identify the nature of damage which will be discussed elsewhere in this book. Damage zone or process zone in PE consists of thin fibers and membranes drawn from the original homogeneous material. Under fatigue and creep loading, crack initiation is an ultimate outcome of a slow process of damage accumulation on sub-micro and microscales. The process progresses faster in the vicinity of preexisting micro-defects such as cavities, gels and/or foreign particles, agglomeration of additives resulting from poor mixing, and possibly others. For an existing field of material imperfections, the crack-initiation time strongly depends on service (or test) conditions such as the stress level, strain rate, temperature, and potentially an aggressive environment. A large scatter, commonly observed in PE lifetime, is primarily associated with the random geometry, chemical composition, and the location of the defect, triggering crack initiation. In contrast, the crack in degradation-driven brittle fracture does not require the presence of a defect [Choi et al. (2005, 2009)]. A thin surface layer of PE component that is in direct contact with a chemically aggressive environment undergoes degradation. Chemical degradation of semi-crystalline PE is manifested in a reduction of molecular weight (MW) that often leads to an increase of crystallinity and therefore PE density. The densification of the surface layer attached to the interior unchanged material results in a buildup of residual stress: Tensile in the degraded surface layer and compressive in the interior domain. The degradation of surface layer also leads to material embrittlement, i.e., dramatic reduction of PE toughness. As the degradation progresses, a combination of increasing tensile stress and decreasing toughness of degraded PE leads to multiple crack initiation within the degrading surface layer [Choi et al. (2005, 2009)]. Although a defect is not necessary for ultimate failure, the presence of a defect within the surface layer may accelerate crack initiation reminiscent of comorbidity when a human body is invaded by virus. Now let us delve into crack propagation in PE followed by initiation.

Cracks with surrounding process zones (i.e., CL) are commonly observed in PE. Similar to other materials, the process zone in polyethylene consists of active zone moving in front of the crack and wake zone left behind, when crack advances through the active zone. The AZ has a narrow wedge shape extended in front of the crack along the crack line and consists of thin fibers and membranes drawn from the original homogeneous material. There is a sharp boundary separating PZ (drawn material) from the surrounding original material. Figure 5.31 shows a general view of CL emanating from a notch (on the left). Active zone here is barely visible as the light gray part of CL triangle toward the tip. The crack propagation is the process of breaking fibers under creep condition and forming irregular fracture "surfaces" consisting of the broken fibers. Such "surface" is very different from the conventional Fracture mechanics model of crack as a cut. A part of the wake zone, marked on the general view by a rectangular box, is shown with higher magnification on the lower micrograph. Note a very sharp boundary that separates the process zone from the original material. The fibers are formed by cavitation followed by so-called cold drawing process. The cold drawing takes place at the draw stress σ_{dr} that depends on the drawing rate. The stress acts on AZ from the specimen side along the boundary of quasi-equilibrium coexistence of drawn and original materials. The draw ratio λ, i.e., the ratio of the length of drawn fiber and the thickness of the original material, is usually in the range 4–6 (depending on the PE grade). There is a well-defined conical shape of the fibers, which suggests fiber creep-related fracture. The width of the original material that is transformed into fibers is uniquely determined by (a) external load; (b) material parameters such as the drawing stress σ_{dr} and constant draw ratio λ; and (c) two independent CL parameters: The crack length l^{CR} and the active zone length l^{AZ}. A part of the PZ is referred to as active zone (AZ), since there are active processes of cold drawing and creep of oriented material within it. The AZ has a narrow wedge shape and extends along the crack plane (Figure 5.31). It consists of thin microfibers and membranes drawn from the original homogeneous material. Creep and ultimate rupture of drawn microfibers within the AZ results in crack extension through the mid-plane of the AZ and forms a crack with "hairy" surfaces. The integral effect of energy absorption by cold drawing of microfibers and membranes within the AZ and by creep of the drawn material prior to its ultimate rupture is the origin of PE SCG resistance: i.e., fracture toughness. Thus, the crack growth is closely coupled with formation and evolution of the AZ in front of the crack. A coupled system of the crack and the AZ is referred to as the Crack layer (CL) [Chudnovsky, A. (1984, 2014)]. A typical appearance of a CL in PE is shown in Figure 5.31 [Chudnovsky, A. (2014)]. In Figure 5.31, a relatively white and gray domain of length l^{cr} with well- visible ruptured fibers shows a crack (displacement discontinuity) with "hairy" upper and lower surfaces. There are sharp boundaries that separate the conically shaped cold drawn PE fibers from the original material (the dark domains above and below the crack). The crack front is easily identified by a sharp increase of the crack opening in comparison to that within the AZ, when oriented PE fibers bridge the upper and lower faces along the AZ length l^{AZ}. The total CL length is $L = l^{cr} + l^{AZ}$. The AZ width is not an independent parameter, since it is determined by (a) PE natural draw ratio λ_n and (b) by the displacement field computed along the boundary of the entire CL domain cutoff [Chudnovsky et al. (2012)]. Thus, l^{cr} and l^{AZ} are only geometric characteristics of CL. The rupture of the oriented PE within the AZ results in the crack advance into AZ and therefore reduction in AZ length. In response, the AZ growth away from the crack takes place increasing the AZ length. AZ width evolves in time: $W_0 = W_0(x,t)$ where W_0 is a function of location x within AZ and time t. The width of AZ in actual configuration (after drawing) is $W_0 = \lambda W_0$ (λ is the natural draw ratio). Thus, the displacement discontinuity normal to the crack direction due to cold drawing is $(\lambda - 1)W_0$. To determine the amount of material undergoing cold drawing and being incorporated into AZ, i.e., the width W_0, we consider crack opening displacement (COD) as an approximation of the opening of a slit in the specimen with PZ cutoff. The approximation is quite accurate due to the presence of a small parameter: The ratio of the width over the length of the slit. Then, we make use of conventional Fracture mechanics formalism and compute COD for the boundary value problem. Thus, W_0 can be found from the continuity conditions of the original problem: the COD, $\delta(x,t)$ in the specimen with PZ cutoff, and the traction σ_{dr} along AZ boundary should be equal to the displacement discontinuity in drawing process occurring within the AZ: $W_0(x,t) = (\lambda - 1)^{-1} \delta(x,t)$. The COD depends on applied load σ_∞, σ_{dr}, the specimen dimension W as well as crack length l^{cr}, AZ length l^{AZ}, and CL lengths $L(L = l^{cr} + l^{AZ})$. Thus, in this case, W_0 is not an independent CL parameter. That is the reason for the simplification of CL model in application to PE and its reduction to two-parametric model. The stationary (quasi-equilibrium) state of a solid with two-parametric CL is conventionally determined by the minimum of the total Gibbs potential of the system. The total Gibbs potential is the sum of the Gibbs potentials of the specimen with PZ cutoff G_0 and the PZ potential G_{PZ}. The Gibbs potential of PZ in turn consists of the wake zone G_{WZ} and active zone G_{AZ} potentials. AZ consists of homogeneous drawn material with properties different from that of the surrounding original material. It is also separated from the original by distinct boundary with drawing stress acting along it. As a result, the AZ Gibbs potential can be presented as $G_{AZ} = \gamma^{tr}.V_{AZ}$ where the multiplier γ^{tr} stands for the specific energy of cold drawing γ_0^{tr}, i.e., the work required to transform a unit mass of original material into an equal mass of oriented unstressed state plus the difference of strain energy densities in the original and drawn states, assuming both are linear elastic materials (Eshelby, 1951): $\gamma^{tr} = \gamma_0^{tr}(T) + \dfrac{\sigma_{dr}^2}{2}[\uparrow C]$. V_{AZ} is the volume of the material in AZ, and $\uparrow C$ is a jump of material compliance across the active zone boundary. The potential of the wake zone, G_{WZ}, is different in two ways: (a) It does not contain the strain energy term, since the crack is traction free, and (b) it contains the surface energy term of the crack faces. The volume V_{AZ} of AZ material in the reference state can be obtained by integrating the AZ width, which is expressed in terms of total COD in the specimen with PZ cutoff.

Thus, CL propagation takes the form of the AZ and crack-tip movements in material coordinates. The driving forces for such processes are defined following the thermodynamic formalism mentioned earlier [Glansdorff, P. and Prigogine, I. (1971)], and they are expressed as the derivative of Gibbs potential with respect to l^{AZ} and l^{cr} respectively: The AZ driving force $X^{AZ} \cong -\partial G / \partial l^{AZ}$, and the crack driving force $X^{cr} \cong -\partial G / \partial l^{cr}$ [Chudnovsky, A. (1984, 2014)]. Both driving forces have similar expressions:

$$X^{AZ} = J_1^{AZ} - \frac{\gamma^{tr}}{\lambda - 1} \delta(x) \big|_{x=l_{cr}} \tag{5.75}$$

where J_1^{AZ} is the elastic energy release rate associated with the AZ growth into the original material PE domain, δ stands for displacement discontinuity within CL, and $\delta(x)/\lambda - 1 \big|_{x=l_{cr}}$ is the thickness of the original PE from which the AZ was drawn at the crack tip. The crack driving force has a similar expression $X^{cr} = J_1^{cr} - 2\gamma$. Despite the similarity of the two expressions, X^{AZ} and X^{cr} are essentially different. J_1^{cr} is the elastic energy release due to the unit crack advancement into the AZ also presented by a path-independent integral with an integration path strictly confined within AZ. The γ^{tr} in X^{AZ} stands for the specific energy of the original PE transformation into the drawn, oriented PE, in contrast with 2γ which represents the work done on oriented PE rupture under creep.

Conventionally, the SCG kinetic equation is formulated as an empirical relation between crack growth rate l^{cr} and SIF, or, in more general case, ERR, G_1, J_1^{cr} [Chudnovsky, A. (2014) Grellman, W. and Seidler, S. (Eds.) (2001), Chan, M.K. and Williams, J.G. (1983), Paris, P.C. and Erdogan, F.A. (1963)]. However, in the case of the CL system in PE, we have two dynamic variables:

The crack and AZ lengths l^{cr} and l^{AZ}. Therefore, the kinetic equations of fracture propagation imply functional relations between the crack and AZ growth rates $\dot{l}^{cr}$ and $\dot{l}^{AZ}$ on one side and corresponding driving forces X^{cr} and X^{AZ} on the other side. Adapting classical Onsager's arguments for the proportionality between thermodynamic fluxes and corresponding forces for a relatively small deviation from equilibrium and considering the irreversibility of crack and AZ growth (fluxes), the following system of CL equations have been studied [Eq. (5.76)] [Chudnovsky, A. (2007)].

Thus, the role of AZ is in moderating high stress concentration caused by an inclusion or a crack. It is achieved by strain localization (displacement discontinuity) in the form of cavitation and cold drawing. However, AZ can only delay the crack growth process: The creep and/or other types of degradation of the AZ material reduce its toughness with time and ultimately allow crack growth into AZ. Such process of CL propagation continues by crack and AZ assisting mutual growth. Thus, it is important for SCG model that CL in PE is a system with two degrees of freedom. The described scenarios of SCG are formalized in the following system of coupled ordinary differential equations with respect to crack and AZ:

$$\left\{ \dot{l}^{cr} = k_1 X^{cr} \left(l^{cr}, l^{AZ} \right) \text{ if } X^{cr} \geq 0, \text{and } \dot{l}^{cr} = 0, \text{if } X^{cr} < 0 \right.$$

$$\left\{ \dot{l}^{AZ} = k_2 X^{AZ} \left(l^{cr}, l^{AZ} \right) \text{ if } X^{AZ} \geq 0, \text{and } \dot{l}^{AZ} = 0, \text{if } X^{AZ} < 0 \right. \tag{5.76}$$

The kinetic equations Eq. (5.76) are supplemented by initial conditions $l^{cr}(0) = l_0^{cr}$, $l^{AZ}(0) = l_0^{AZ}$, and a law of fiber creep that determines decay with time of the specific fracture energy $2\gamma(t)$. The CL driving forces X^{cr} and X^{AZ} are nonlinear functions of crack and active zone lengths and depend on material parameters γ and γ^{tr}. To evaluate such material parameters, it is convenient to use a specimen geometry where ERR does not depend on the variable *crack length*. It also helps to study CL growth mechanisms and kinetics by reducing the number of variables. Tapered double cantilever beam (TDCB) specimens have been used for crack growth studies in metals [Mostovoy et al. (1967)] for this purpose. However, a TDCB specimen made of PE displays a very large deformation that compromises the accuracy of basic fracture parameters' evaluation. Another specimen geometry, a stiff constant K (SCK) specimen more rigid than TDCB and thus more suitable for studies of SCG in PE, has been developed [Chudnovsky et al. (2012)]. Care must be taken for SCK specimen, since SIF is independent of crack length only in a certain range of crack length.

The crack and AZ thermodynamic forces are nonlinear functions of crack and AZ lengths. Thus, the system of equations (5.76) despite its simple appearance is a nonlinear system of ordinary differential equations (ODE),

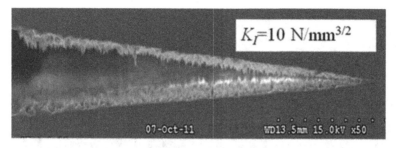

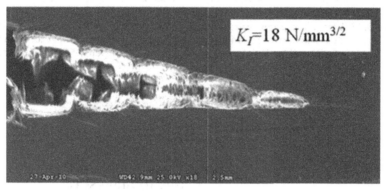

Figure 5.32 CL side views observed in the creep tests conducted on SCK specimens at 80°C, but different constant SIF values are shown on the micrographs. A continuous CL growth (upper micrograph) with a wedge-shaped PZ takes place at a low load level $\left(K_I = 10 N / mm^{1/2}\right)$, whereas discontinuous pulsating PZ is seen at a load 1.8 times higher $\left(K_I = 18 N / mm^{1/2}\right)$.

solution of which calls for numerical methods. An illustration of numerical simulation of CL growth in a single edge-notched (SEN) specimen shows two types of solutions: Discontinuous, stepwise CL growth from one stationary AZ configuration to the next one and continuous CL growth with underdeveloped AZ that never reaches the equilibrium size [Chudnovsky and Shulkin (1999)]. In the stepwise CL growth at the beginning, the crack is stationary, whereas the first AZ grows toward its equilibrium size. The equilibrium size is reached and maintained constant until the degradation of AZ material triggers the crack growth into the AZ, following the first equation of the system (3.12). The crack propagates through AZ and gets arrested again, when it meets the original material at the tip of AZ. The AZ size growth accompanies the crack growth, during this time, and CL reaches the second stationary configuration. A newly drawn material constitutes the new AZ. Then, the same degradation processes take over the newly drawn material, and crack propagates through the second AZ the same way as

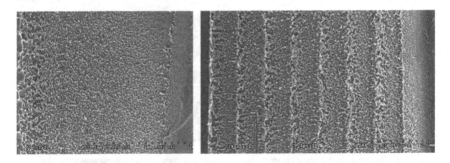

Figure 5.33 Left micrograph represents fracture surface associated with continuous crack growth, and right micrograph represents fracture surface associated with pulsating crack growth represented by striations.

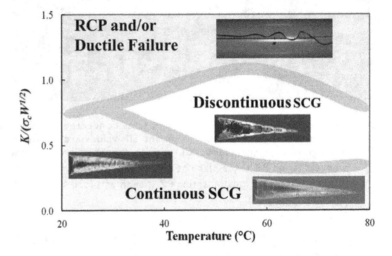

Figure 5.34 Continuous and discontinuous SCG mechanisms and RCP mechanisms displayed in the SIF versus temperature diagram.

it has done it at the first AZ. In single edge-notched specimen (SEN), the maximal value of AZ driving force increases with CL length. As a result, the equilibrium AZ size increases and the duration of steps decreases with step number, leading to an accelerated CL growth and final CL instability and transition to rapid (dynamic) crack propagation leading to final failure. The AZ material degradation time in this example is significantly longer than AZ growth to equilibrium size. As a result, both crack and AZ are stationary most of the time with relatively short time intervals of growth. Another scenario of CL growth is when the AZ material degradation rate is comparable to or faster than the rate of AZ growth. In such case, the crack starts to grow into AZ before AZ reaches equilibrium. An extension of the crack into AZ results in an increase of AZ driving force and an increase of AZ growth rate. Thus, AZ never reaches equilibrium. This process accelerates with an increase of CL size due to SEN geometry, similar to that for a discontinuous growth.

Different mechanisms of SCG in PE, specifically continuous and discontinuous or stepwise SCG, have been consistently observed under constant load and temperature on various specimen geometries [Chudnovsky et al. (2012), Parsons et al. (1999, 2001), Zhou et al. (2013), Zhang, H. and Chudnovsky, A. (2013)]. The different mechanisms of SCG are illustrated in Figure 5.32 by CL side views observed in the creep (constant load) tests conducted on SCK specimens at the same temperature (80°C), but different load levels indicated by constant SIF values are shown on the micrographs [Chudnovsky, A. (2014)]. A continuous CL growth (upper micrograph) with a wedge-shaped PZ takes place at an SIF level $\left(K_I = 10\,N / mm^{1/2}\right)$, whereas discontinuous pulsating PZ is seen at an SIF 1.8 times higher $\left(K_I = 18\,N / mm^{1/2}\right)$.

The corresponding fracture surfaces observed in the same tests are shown in Figure 5.32. In continuous SCG (at low load level), there is a uniformity of fracture surface in the direction perpendicular to the crack growth direction and gradual reduction in the random pit size in the direction of crack growth (left micrograph in Figure 5.33). A discontinuous crack growth depicted by the "pulsating" CL side view (lower micrograph in the first figure, Figure 5.32) is manifested as striations on fracture surface (right micrograph in Figure 5.33). Monotonic and stepwise load point displacement (LPD) versus creep time curves confirm the continuous and discontinuous SCG mechanisms independently on side views and fracture surface observations.

A continuous SCG under creep conditions is expected since there is no variation of load and/or temperature. In contrast, the discontinuous (stepwise) growth pattern is a surprise since there are no external factors. The discontinuous mode of fracture growth is simply a manifestation of the crack and the AZ interaction. A simple mathematical model of such interaction expressed as the aforementioned system of CL growth equations captures the essence of such interaction. The solution of the nonlinear system of equations is not unique and suggests various scenarios of CL evolution such as continuous, discontinuous, and transient from continuous to discontinuous growth. The model predictions are confirmed by experiments and field observations.

Finally, a set of SCG tests conducted with identical SCK specimens led to the formulation of a fracture mechanism map in terms of normalized SIF and temperature, as shown in Figure 5.34 [Zhou et al. (2010, 2013), Zhang, H. and Chudnovsky, A. (2013)]. Fracture Mechanism map for a PE pipe material is shown in Figure 5.34. The fracture mechanism map summarizes the observations illustrated in Figures 5.32 and 5.33 earlier. Specifically, there are domains in SIF–temperature coordinates with different SCG patterns: Continuous SCG for all temperature ranges at low SIF (normalized K_I around 0.5 and below), rapid crack propagation (RCP), or ductile rupture at high SIF (normalized K_I is about 1 and above) also for all temperature ranges. A discontinuous SCG occurs at normalized SIF values in between 0.5 and 0.75 as shown in Figure 5.34. Fracture mechanism map is presented with photos that illustrate SCG patterns. A typical sinusoidal RCP trajectory for pressure pipes is illustrated in the uppermost photo in Figure 5.34. Continuous and discontinuous SCG mechanisms are closely associated with different SCG kinetics: Continuous propagation has a much higher crack acceleration than for a discontinuous growth shortening the PE lifetime.

The life expectancy of thermoplastics in durable applications varies from about 10 to 50 and even 100 years in certain cases. It calls for an accelerated testing of material and structures. The challenges of accelerated testing for lifetime are (a) to reproduce the mechanisms of field failures and (b) to develop a reliable procedure for the extrapolation of a relatively short test data into long-term service conditions. Acceleration of fracture by high stress level turns to be inadequate, since the fracture mechanisms change with stress level. Thus, the application of existing standards such as ASTM F1473, ASTM 2837, and ISO 9080 for the estimation of PE lifetime at ambient temperature on the basis of elevated temperature testing at certain SIF level where there is a different mode of SCG may lead to gross error in the prediction of service time. A discontinuous, stepwise SCG has been consistently observed at elevated temperature (80°C and 60°C) in contrast with commonly observed continuous fracture propagation at 23°C. The different mechanisms and kinetics of SCG are controlled at the microscale by creep deformation and fracture of the cold drawn fibers within the AZ. Continuous SCG results from the rupture of microfibers/membranes of AZ during or shortly after the completion of necking. Rupture of the AZ material is the mechanism of crack growth in PE. In contrast, when the fibers/membranes survive the cold drawing process and reach a quasi-equilibrium coexistence with the surrounding original PE, a creep of drawn material may take a sufficiently long time before fracture (the creep strain rate of drawn PE is significantly lower than the original). During creep of drawn material, the PZ size remains almost the same. Then, after a certain time, microfibers/membranes in the drawn material break, leading to a spontaneous crack advance almost to the end of the AZ. Thus, the creep of drawn material within the AZ is the origin of the stepwise, discontinuous SCG in PE.

Recognizing that continuous versus discontinuous SCG patterns are determined by brittle versus ductile failure of microfibers within the PZ, the continuous–discontinuous SCG transition has been referred to as ductile–brittle transition (DBT2) in impact toughness [Zhou et al. (2010), Chudnovsky, A. (2014)]. The corresponding DBT2 temperature depends on SIF. The relation between SIF (representing applied load) and (transition temperature) TT2 is shown by the lower boundary of the discontinuous SCG domain on the fracture mechanism map [Zhou et al. (2013), Zhang, H. and Chudnovsky, A. (2013)]. In summary, a transition in the mechanism and kinetics of SCG, DBT2, presents a limitation for the employment of an elevated temperature as an accelerating parameter for brittle fracture. An extrapolation of SCG data from one domain of fracture mechanism map into another across the DBT2 boundary is not acceptable since it violates the similarity of SCG processes. Therefore, the commonly used acceleration technique based on extrapolation of 1–1.5 years long test data at elevated temperatures into 50-year or 100-year lifespan in service condition is a suspect. However, there is no evidence for using PE for 100 years to validate these predictions. We also know that engineering PE grades in recent years is much improved from past, and old service lifetime data will provide no validation for current PE-accelerated test predictions. In addition, there is a silent assumption that after decades of service, PE retains its original properties tested during qualification. The material aging process (chemical degradation and/or physical aging) commonly takes place over time and should be taken into consideration. It seems the "empirical" extrapolation method works well as a marketing tool rather than an engineering method of lifetime assessment. A molecular characteristic like "tie chains" could be an equally viable marketing tool to compare PE materials with better ability to differentiate [Patel et al., May 2, 1996; Huang, Y.-L. and Brown, N. (1988)].

The acceleration of testing for lifetime by elevated temperature is the most widely used technique at the present. This paradigm, however, faces a problem associated with the changes in the mechanism and kinetics of slow crack growth (SCG). At a certain combination of load and temperature, a transition from a continuous SCG to discontinuous, stepwise crack propagation has been recorded. Optical and scanning electron microscopy observations suggest that the change of SCG mechanisms is closely related to the material ability to form in front of the growing crack a stable process zone that consists of single or multiple crazes and/or shear bands. The crack acceleration in

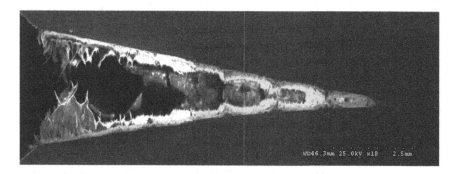

Figure 5.35 Side view of CL developed at 80°C.

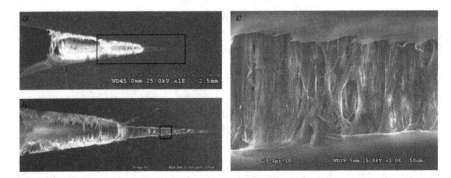

Figure 5.36 (a) Side view of CL developed at 60°C, (b) close view of the boxed area in (a), and (c) close view of boxed area on (b).

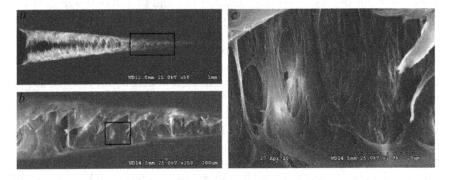

Figure 5.37 (a) Side view of CL at 23°C (SEM, 50×); (b) close view of AZ with unbroken fibers (SEM, 250×); and (c) high-magnification (2000×) micrograph of highly oriented fibers within AZ.

the continuous growth mode is observed to be significantly higher than that in stepwise propagation. Such changes in the mechanism and kinetics of SCG are associated with a transition from a ductile to brittle behavior of microfibers within the process zone. It is referred to as ductile–brittle transition of the second kind (DBT2) based on a resemblance with well-known ductile–brittle transition in dynamic impact resistance. DBT2 is presented in form of SCG mechanisms map in temperature–stress intensity factor coordinates. SCG mechanism map implies certain limitations for the extrapolation of conventional temperature accelerated test data to the service conditions of plastic components. An alternative to conventional accelerated testing approach to evaluate lifetime of plastics structures is proposed in this chapter. It consists of three steps. The first is a characterization of the defects population that may be responsible for fracture initiation. The formulation of constitutive

Figure 5.38 A close view (500× magnification) of highly oriented and broken fibers within WZ at 23 °C.

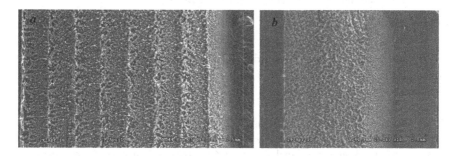

Figure 5.39 (a) Typical appearance of the fracture surfaces formed in creep test at 80 °C with regularly spaced striations and (b) statistically homogeneous surface with no striations at 23 °C (SEM, 18×).

equations of fracture process based on specially designed tests is the second step. Numerical simulation of fracture process using constitutive equations developed within the second step and the evaluation of the lifetime of plastic structure comprise the third step. A validation testing of the proposed program is required.

Continuous versus discontinuous mechanism of SCG is revealed by CL side view, observations of fracture surfaces, and the records of the load point displacement (LPD). The side views of CL observed at elevated temperatures are displayed on the Figures 5.35 and 5.36. Figure 5.35 shows a general view of CL as it appears on the central cross section of the specimen along the vertical plane of symmetry. The CL was formed after 149 h of creep under SIF K = 22.9 MPa.mm$^{1/2}$ at 80 °C. A white, wedge-type pulsating domain displays a discontinuously grown CL with six well visible steps. Five pulsations (steps) are filled with drawn and broken fibers. The sixth one, a weakly visible and elongated triangular domain, shows AZ with freshly drawn intact fibers. The

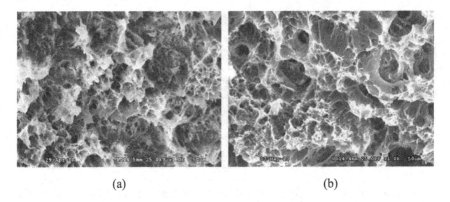

(a)   (b)

Figure 5.40 Close view of fractured newly formed AZ with visible cavitation and drawing of material between the cavities.

fibrillated (white) PZ is surrounded by a uniform dark grey domain of the original PE. There is a sharp boundary separating PZ from the original PE. It is also interesting to notice that all five completed steps (pulses) are the same length.

A similar CL developed during 572 h of creep under SIF K = 25.1 MPa. mm$^{1/2}$ at 60°C is depicted in Figure 5.36a. There are only two completed steps with broken fibers and intact AZ in front. The boxed area encompasses the second step of the WZ with broken fibers and AZ, which can be seen in front. The higher magnification (100×) of that boxed area is shown in Figure 5.36b. Here, the opening near the tip of WZ is about 600 μm with broken fibers of conic shape with about 20–50 μm diameter at the base and up to 150 μm in height. There is a sharp narrowing down in transition from WZ with broken fibers to AZ that has an elongated triangular shape with opening at the root of about 150 μm. There are also two pairs of weakly visible symmetrical "flankers" emanating from the tip of WZ on both sides of AZ. Notice a conic shape of broken fibers, which is discussed later. Figure 5.37c shows higher magnification (2,000×) of the boxed area in Figure 5.37b. At such a magnification, one can clearly see highly oriented dendroid-type fibrillated texture of about a few microns in size within AZ and smooth, homogeneous original PE on both sides of AZ. Notice a sharp boundary that separates AZ material from the surrounding original PE. The fibers of AZ are stressed by the load transmitted by the original material and experience creep. In addition, a fresh original material is being drawn into AZ, and the boundary moves away from the plane of AZ symmetry. A combination of creep and drawing of additional material leads to coarsening of the fibers prior to break. Figure 5.37a presents a general view of CL developed during 1,602 h of room temperature (23°C) creep under SIF

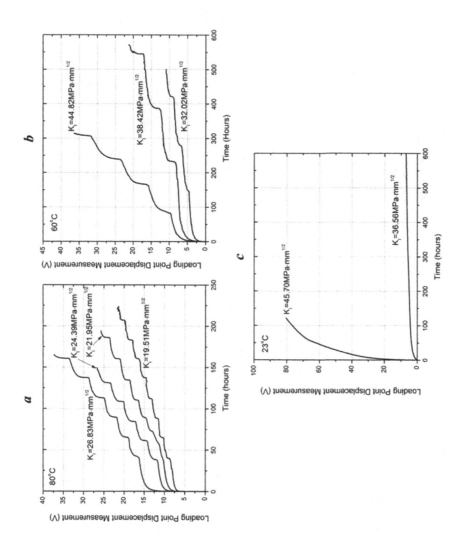

Figure 5.41 LPD recorded in creep test conducted with various SIFs at (a) 80°C, (b) 60°C, and (c) 23°C.

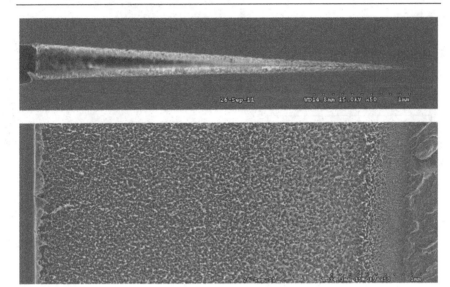

Figure 5.42 Side view and fracture surface of CL grown at 80°C under relatively low SIF (K = 10 MPa.mm$^{1/2}$).

K = 36 MPa.mm$^{1/2}$. There are no steps or pulsations in contrast with CL developed at elevated temperatures, rather monotonically decreasing width of the triangular CL toward a very weakly visible tip. A close view of the tapered shape near tip domain of AZ with an opening of about 120 µm on the left and 40 µm on the right side is shown at 250× in Figure 37b. The last, high magnification micrograph (2,000×) shows a highly oriented wall with µ-dendroid-type fibrillated texture of a few µm diameter, similar to what we have seen at 80°C and 60°C tests. A close view of broken fibers within WZ formed in CL growth under creep at 23°C is depicted in Figure 5.38. Notice the conic shape of broken fibers similar to that observed at elevated temperatures. Another clear evidence of discontinuous versus continuous CL growth at elevated and room temperatures are presented in micrographs of fracture surfaces. Figure 5.39a shows a typical appearance of fracture surface resulting from CL growth under creep at 80°C. The pulsations observed in the side view also make clear marks in the form of striations. The striations are equally spaced, suggesting the same size of AZ at each step. The very last step on the right side of the micrograph in Figure 5.40a corresponds to AZ that has been fractured at the liquid nitrogen temperature to preserve the AZ structure. Due to the discontinuous crack growth through formation of AZ followed by rupturing through it, it leads to the formation of striations. It has a different appearance from all previous steps. Figure 5.40b displays a typical fracture surface developed during the creep test at room temperature.

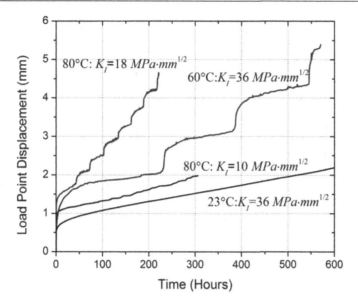

Figure 5.43 LPD at various temperatures and SIFs in SCK specimens.

It shows a homogeneous fibrillated surface with no evidence of any striations. At the very end, we can see a smooth strip that corresponds to the last AZ broken in liquid nitrogen similar to that observed in Figure 5.40a. High-magnification (2,000×) SEM micrographs of fracture surface displayed in Figure 5.40 reveal the micro-mechanism of AZ formation. First, we can see cavities of various sizes in the range between 1 μm and 10 μm randomly distributed in the domains of observation. The cavities are surrounded by very thin membranes that constitute the walls within AZ with dendroid-type texture we observed in Figures 5.38 and 5.39.

One more piece of evidence of discontinuous versus continuous CL growth is presented in Figure 5.41 in the form of load point displacement (LPD) versus time records monitored in the creep tests at various loads (SIFs) and temperatures. Figure 5.41a displays graph of LVDT output corresponding to LPD measured at room temperature, 60°C, and 80°C. The SIF values corresponding to LPD measurements are also marked in Figure 5.41. The LPD graphs at 60°C and 80°C tests clearly reveal the stepwise SCG process. Almost constant duration of the individual steps can be seen independent of SIF for each temperature. In contrast with that, there are no visible steps in LPD record at room temperature (Figure 5.41c). The only information available here is the total duration of the test. It should be noticed that the duration of the very first step in the stepwise SCG is quite different from the well reproducible duration of the following steps. The first step is associated with the formation of the very first AZ from the artificial notch made to initiate

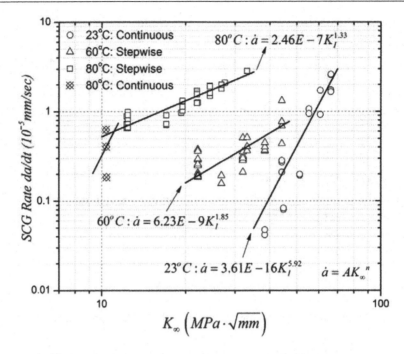

Figure 5.44 Average SCG rate as a function of SIF in form of Paris–Erdogan equation.

the fracture. The tip of the AZ is much sharper than the notch. In general, crack-initiation process in PE is associated with formation and fracture of the first AZ. Therefore, the understanding of CL growth sheds light on fracture initiation as well. The data presented earlier at first suggested that the transition from continuous to discontinuous SCG is associated with temperature. However, further studies reveal that continuous growth takes place at elevated temperature as well, but under low enough load (SIF). For example, Figure 5.42 depicts the side view and fracture surface of CL developed at 80°C and relatively low SIF $K = 10\,\mathrm{MPa.mm}^{1/2}$.

The appearances of CL side view and fracture surface in Figure 5.42 are similar to those observed at 23°C (see Figures 5.38, 5.39, etc.): No sign of discontinuous growth. Clearly, continuous SCG occurs at elevated temperatures within a certain range of load (SIF). Collaborating evidence of continuous versus discontinuous SCG is provided by the LPD record. Figure 5.43 displays four LPD versus time curves from four different tests. The graphs labeled 80 C-1 and 80 C-2 corresponds to 80°C testing with relatively high $\left(K = 18\,\mathrm{MPa.mm}^{1/2}\right)$ and relatively low $\left(K = 10\,\mathrm{MPa.mm}^{1/2}\right)$ SIF. Continuous and discontinuous SCG for low and high SIFs are well manifested in the

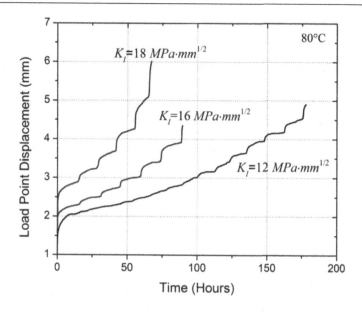

Figure 5.45 LPD at 80°C for various SIFs in CT specimens.

monotonous and stepwise shape of 80 C-2 and 80 C-1, respectively. Two other graphs labeled 60°C and 23°C correspond to two tests conducted at the same SIF $\left(K = 36\,MPa.mm^{1/2}\right)$, but different temperatures (60°C and 23°C, respectively). It suggests that $K = 36\,MPa.mm^{1/2}$ is high enough SIF for 60°C to cause discontinuous SCG and is relatively low for 23°C resulting in continuous SCG. Apparently, SIF should be properly normalized in order to compare loading at different temperatures.

Combining fractographic measurements with LPD records, we evaluate SCG rate for different loads and temperatures. Specifically, the average crack-growth rate for a discontinuous growth is evaluated as the ratio of the average distance between striations measured from the fracture surface micrographs and the average individual steps durations measured from LPD versus time record. A set of statistically valid average crack-growth rate is obtained from one test due to constancy of SIF and in most cases sufficient number of reproducible steps in one specimen. For continuous SCG, the average crack-growth rate is determined as the ratio of total crack extension during SCG measured from the fracture surface and the time interval between the beginning and the end of the test. Thus, a multiple specimen test is required to obtain statistically valid crack growth rate for continuous SCG. Such tests usually result in higher uncertainty of the data. Figure 5.44 shows in a double logarithmic scale, the statistical data and the best fit for average crack growth rates ($\dot{a}$ in 10^{-5} mm/second versus SIFs K_{∞} in MPa.mm$^{1/2}$ for

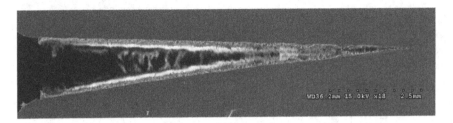

Figure 5.46 CL side view (CT specimen, initial KI0 = 11.5 MPa.mm$^{1/2}$, a0 = 7.6 mm, 80°C).

three temperatures of 23°C, 60°C and 80°C). Conventional Paris–Erdogan equation (power law) for crack growth is used to treat the data:

$$\frac{da}{dt} = A \cdot K_1^n \left(\sigma, a, a / W\right) \tag{5.77}$$

It can be seen from Figure 5.44 that both parameters A and n in Paris–Erdogan equation are noticeably different for continuous and discontinuous crack growth. Specifically, the values of power n for the discontinuous growth at 60°C and 80°C are close. However, for the continuous growth, the power n is about three times larger than that of discontinuous growth. It should be noticed that the continuous growth data should be considered as preliminary due to a very high level of uncertainty in small number of experiments with continuous growth at elevated temperature. Reliable data are obtained for the room temperature (23°C).

Summary of Observations and Discussion

AZ in PE appears as a narrow wage-type strip of highly oriented (cold drawn) membranes and fibers with a sharp boundary between domains of oriented and original isotropic material for all temperatures. A specially designed specimen with SIF independent of crack length employed for SCG studies provides well-reproducible data and does not require monitoring crack length for the evaluation of crack growth rate. Different mechanisms and kinetics of SCG, i.e., discontinuous growth with relatively low acceleration and continuous growth with significantly higher acceleration are observed. Such transition in crack growth pattern in general depends on load level, crack size, specimen geometry, and test temperature. The load and crack-specimen configuration can be represented by one scaling parameter: SIF. It suggests that at a given temperature, the transition may take place in process of crack growth in a specimen with increasing SIF with crack length [Chudnovsky et al. (2012)]. Such proposition has been

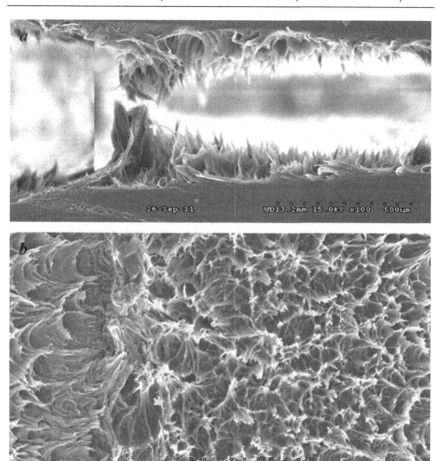

Figure 5.47 (a) Side view and (b) top view of striation domain (CT specimen, 80°C, initial K = 19 MPa.mm$^{1/2}$).

examined using compact tension (CT) specimen with the same dimensions as SCK, but without holes. The tests were conducted at 80°C. Figure 5.45 displays the CT specimen LPD curves for three different load levels represented by the initial SIF: high load CT1 (initial SIF $K_{10} = 18$ MPa.mm$^{1/2}$), intermediate load CT3 (initial SIF $K_{10} = 16$ MPa.mm$^{1/2}$), and low load CT6 (initial SIF $K_{10} = 11.5$ MPa.mm$^{1/2}$). The CT1 and CT3 curves have a well-pronounced stepwise pattern corresponding to a discontinuous crack growth. CT6 curve is relatively smooth at the beginning, but after about

100 h of growth, i.e., with increase of crack size and SIF, the stepwise pattern starts to emerge. CL side view for CT6 specimen is depicted in Figure 5.46. It corroborates with LPD data, showing well-visible PZ "pulsation" at advanced crack and no visible evidence of discontinuous growth at the beginning. Thus, the transition in CL growth pattern at a given temperature depends on SIF. A close look at an individual striation formed in CT specimen at 80°C under initial $K = 19 \, MPa.mm^{1/2}$ is presented in Figure 5.47. It shows a higher magnification (100×) of the side view (Figure 5.47a) and top view (Figure 47b) of the striation domain. Figure 5.47a displays a large bundle of drawn fibers forming a ligament that separates the crack from a set of much smaller fibers in front of it. The ligament size correlates with Crack Mouth Opening Displacement (CMOD). All broken fibers have conic shape suggestive of creep failure. Thick bundle of fibers (ligament) is noticeably larger than thin individual fibers on the right side of it. It suggests that the small fibers have been broken while the ligament was still intact undergoing creep, and the duration of the stationary AZ in stepwise CL growth depends on the lifetime of the ligament. The reported observations raise a number of questions such as what factors control the fracture propagation pattern, what controls the stationary AZ size (the distance between striations on fracture surface), how to determine the duration of the steps in discontinuous CL growth, how to connect the observed crack with conventional Fracture mechanics parameters, and many others. We notice that AZ growth from the very beginning is closely associated with the transformation of the original homogeneous material into highly oriented (drawn) discrete fibers and/or membranes. Then, the creep of the fibers leads to fiber fracture and transition from AZ into WZ. An example of this type of data analysis can be found in works of Chudnovsky and Shulkin (1999) and Choi et al. (2009). Thus, the numerical simulation of CL growth opens for a new interpretation of a large body of experimental data accumulated over decades of experimentations with SCG under fatigue and creep conditions commonly presented in the form of Paris–Erdogan equation.

Cold Drawing Phenomenon in Semi-crystalline Thermoplastics

An alternative approach to PE lifetime assessment is based on (a) a fundamental understanding of underlying mechanisms of PE deformation and fracture, (b) the experimental evaluation of basic material parameters determining PE behavior, and (c) a sound physical model for the time to fracture (lifetime). For example, understanding of the relation between delayed necking and PE pressure pipe ductile failure combined with a symmetry between yield stress and draw stress dependency on the logarithm of strain rate on one side and the logarithm of failure time dependency of applied stress on the other leads

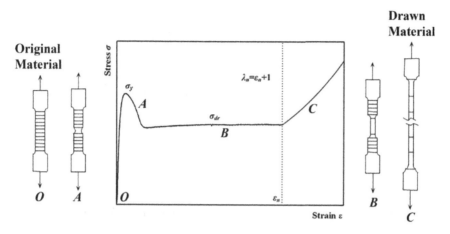

Figure 5.48 Schematics of stress–strain curve for uniaxial tension test and specimen configurations at various stages of the test.

to a simple predictive model for PE lifetime in ductile failure using a relatively short-range testing program. Similarly, a fundamental understanding of the connection of creep of original and drawn PE with mechanisms and kinetics of fracture propagation in PE presented in the form of a system of equations for crack and AZ leads to a numerical simulation of fracture propagation and prediction of lifetime in brittle failure [Zhang et al. (2014)]. Thus, a set of tests for experimental determination of basic PE properties combined with a numerical modelling of ductile and brittle fracture substitute for the empirical procedures of ISO9080 and other similar standards. It is important to keep in mind that for more accurate predictions, PE-aging process needs to be considered. Some very long-term aging data can be included in numerical modelling to account for various service conditions.

Thus, CL growth process is strongly related with the transformation of an isotropic original material into a highly oriented one. Such a transformation is well known in polymers and is called "cold drawing" or "necking". Engineering stress–strain curves recorded in a simple uniaxial tension test have been widely used to study thermomechanical properties of polymers. Figure 5.49 presents a schematic illustration of the typical stress–strain curve observed in such tests with semi-crystalline polymers like PE (pipe grade HDPE) at room temperature (23°C). An illustration of the test specimen configurations on both sides of the stress–strain curve corresponding to the different stages of the test is presented here. On the left side, there is a sketch of an original specimen. The horizontal marks have been printed on the specimen to monitor the deformation in material (Lagrangian) coordinates. At the very beginning of the curves (~2% strain), the process of deformation

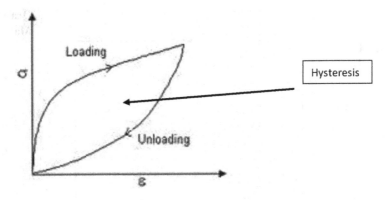

Figure 5.49 Schematic of loading and unloading leading to hysteresis.

is almost linear elastic (elastic deformation implies that the process is fully reversible and that the unloading follows the same path as the active loading). In this region, the elastic modulus can be obtained from the stress–strain curve in the conventional way. As the deformation increases, the nonlinearity of the curve becomes more and more pronounced, eventually approaching the yield point as shown in Figure 5.49. If the loading stops before reaching the yield point and the specimen is unloaded, the unloading path may return to the origin but departs from the initial loading path thus forming a so-called hysteresis loop (Figure 5.49). The area of the loop in stress–strain coordinates presents the energy dissipation and suggests that the loading–unloading process may be mechanically reversible, but thermodynamically irreversible; i.e., during hysteresis, some irreversible changes of PE molecular architecture and/or morphology occur. At the yield stress, the crystal structure is fragmented and partially melted, which allows rearrangements of molecular and crystalline. Randomly coiled molecular chains between the crystals as well as some of the macromolecules initially incorporated into crystals become highly stretched and oriented. Then some of the oriented molecules are incorporated into newly formed crystals and "freeze" the newly formed oriented state. The stretching in axial direction is accompanied by thinning (reduction of cross-sectional area) since the volumetric change is very small if any. Such strain localization takes place at a random location within the specimen, as shown on the second from the left sketch of the specimen in Figure 5.48. The strain localization occurs much faster than the crosshead speed of the testing machine. As a result, an unloading is recorded with the stress reduction from yield stress σ_y to the drawing stress σ_{dr}. The fully developed strain localization ("neck") of the material between two neighboring marks on the original specimen is depicted on the first sketch of the specimen on the right side from the stress–strain curve (Figure 5.48). The process of such strain localization is called "cold drawing" or shortly "necking" (later, we use both terms interchangeably). Notice the difference in the distance between the same two marks in the initial and drawn state. The oriented (drawn) state formed in the necking process is preserved upon unloading of the specimen. The ratio of the distances between the same marks in the drawn and initial states is called "natural draw ratio" λ_n. The recorded increase of the gage length results exclusively from the increase in λ_n times the drawn domain length in comparison to the length of the corresponding original PE. Thus, the movement of the neck boundaries is λ_{n-1} times smaller the displacement at the grips. Following yield point, the engineering stress remains almost constant with continuing increase of strain (Figure 5.48). Once the neck is developed during yield, it spreads across the specimen length without premature failure, the plateau ends, and the stress starts to rise again with increasing strain until the ultimate fracture takes place. After initiation of necking, observed plateau on stress–strain curve with a practically constant level of stress σ_{dr} resembles metal plasticity. However, it is a deceptive resemblance since the physics of cold drawing is very different from that of plasticity. During plateau stage, the boundaries of strain localization spread in both directions. The domain of drawn material grows by transforming the initial isotropic material into drawn (oriented) state. No changes in stresses and strains of the initial and drawn materials domain are observed. The observed increase of the gage length results exclusively from the moving boundaries between the drawn and original materials. There is a narrow boundary layer separating the two domains. Mechanical and phase equilibrium across such boundaries is maintained since the transition occurs significantly faster than the boundary movement. When the specimen is fully drawn, the so-called "strain hardening", i.e., increasing stress with growing strain, is observed. It represents the properties of oriented material (see the right sketch of the specimen in Figure 5.48). The properties of initial and oriented states can be extracted after the test by data analysis. Thus, the cold drawing can be viewed as consecutive PE transformations: (a) fragmentation and partial melting of the lamellas of the initial semi-crystalline PE; (b) stretching of the molecular chains, rotating, and orienting the remaining lamella fragments; and (c) crystallization of the oriented molecules and formation of a new highly oriented semi-crystalline state. The first two transformations take place when σ_y of the of material is reached, and the third transformation occurs when both σ_{dr} and critical level of orientation (stretch λ_n) are achieved at the boundaries of the necked region. The drawn (oriented) PE possesses noticeably different mechanical properties (e.g., rigidity, strength, or creep behavior) than the original (isotropic) one. Upon unloading the drawn material, it returns to an external stress-free oriented state. During the entire plateau stage of the necking process, the specimen is composed of two regions with different properties and variable sizes. This region represents a mixture of material properties with a composite specimen behavior dependent on a variable composition of

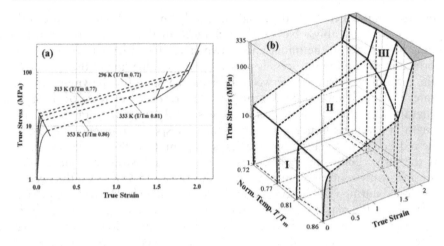

Figure 5.50 True stress–strain–temperature diagram in two-dimensional and three-dimensional forms.

the original and oriented materials. Therefore, the PE properties independent of specimen geometry are reflected by the two parts of the stress–strain curve: One is before yielding and another one is after the cold drawing process is completed, i.e., the strain–hardening region. Thus, engineering stress–strain curve is a complete misrepresentation for a material that undergoes cold drawing, and we must resort to true stress–strain relationships.

In summary, the cold drawing can be viewed as a double transition that happens in three steps. First is the destruction of the semi-crystalline structure of initial material and passing to an amorphous entangled structure. Second is stretching and orienting the molecular chains and remaining small lamella. Third is the partial crystallization of the oriented molecules and formation of highly oriented semi-crystalline material (Mostovoy et al., 1967; Zhou et al., 1995). The first step takes place, when the yield strength of material is reached, and the third step occurs at drawing stress and certain level of orientation (stretching). Naturally, such a double transition is associated with thermal effects related to the latent energy of transformations. The temperature variation was monitored by means of infrared microscope during the drawing process (Mostovoy et al., 1967; Chudnovsky and Shulkin, 1999). It reveals a cooling during thermoelastic stretching of original material and around the yield point where the original crystals are fragmented and/or partially melted, followed by a significant heating during the second crystallization of oriented molecules, which corresponds to drop of stress from σ_y to σ_{dr} on the stress–strain curve (Figure 5.48).

During the entire plateau stage of the process, the specimen is composed of two regions with different properties and variable sizes. Therefore, the

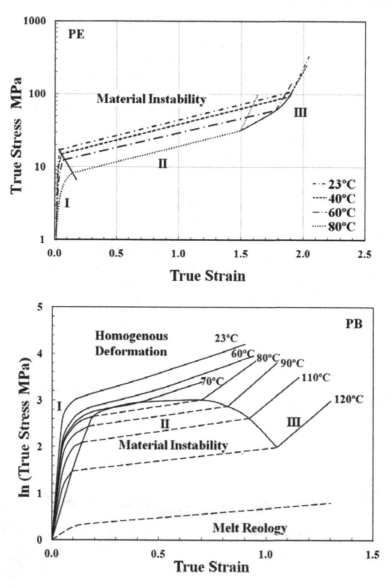

Figure 5.51 True stress–true strain diagrams for PE and PB.

plateau part of the load–displacement curve depicted in Figure 5.48 does not represent material properties but rather a behavior of a composite specimen with variable composition of the original and oriented materials. The true stress versus true strain diagram can be reconstructed from the engineering stress–strain curve (Figure 5.48) using the true stress $\left[\sigma_{tr} = \sigma_{eng}\left(1+\varepsilon^{eng}\right)\right]$ and true strain $\left[\varepsilon^{tr} \cong ln\left(1+\varepsilon^{eng}\right)\right]$ expressions. At small strains, the true stress and engineering stress versus strain curves are very close. However, with increasing strain, the true stress σ_{tr} departs significantly from the engineering stresses σ_{eng}. Figure 5.50 presents true stress–true strain diagram extracted from engineering stress–strain data in 3D space incorporating normalized temperature as the third dimension.

The true elastic behavior of the material before and after cold drawing is presented in the true stress–true strain–temperature diagrams of the two thermoplastics discussed above PE and PB (Figure 5.51). During the entire plateau stage of the process, the specimen is composed of two regions with different properties and variable sizes. Therefore, the plateau part of the load–displacement curve depicted in Figure 5.51 does not represent material properties, but rather a behavior of a composite specimen with variable composition of the original and oriented materials. The diagrams show (Figure 5.51) true stress response to applied true strain $\left(\varepsilon_{tr} = ln\lambda\right)$ of the original material (region I) and oriented material (region III) at different temperatures. The lines of the region II indicate a metastable transition from the original to oriented state, which occurs very fast and is not observable in the test. Notice in PB, necking is observed only at temperatures above 70°C in contrast with PE where necking takes place at all temperatures. Below 70°C, PB displays strain hardening

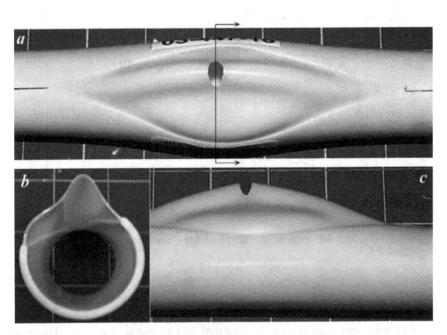

Figure 5.52 External appearance of ductile failure as a ballooning (localized necking).

after yield point without necking. In other terms, there is a clear transition in mechanical behavior of PB around 70°C. It results in significant consequences for crack propagation in PB below and above transition temperature: There is no process zone in PB below 70°C, and as a result crack propagates in classical brittle manner, whereas a process zone (similar to that shown in Figure 5.17) and higher resistance to SCG are observed above transition temperature. The effect of morphology is illustrated in Figure 5.17 by the micrographs of CL developed under fatigue loading at 110°C in PB, a semicrystalline thermoplastic. The specimen was cut from the extruded PB tubing, and as a result the inner surface layer (skin) has a crystalline structure different than that in the core and outer skin due to a different cooling rate inside (slow) and outside (fast). The morphological differences are manifested in different mechanical responses to stresses: Ductile (shear banding) within the inner skin and brittle one (crazing) in the core and outer skin. The cross section a–a displays a complete break and wide crack opening on the outer skin (left side of the lower set of micrographs) and a part of the core in contrast with intact inner skin and an adjacent part of the core with large deformation. Thus, the crack front extends farther on the outer side of PZ than on the inner one. Large white area on the optical micrograph represents a very intensive cavitation, crazing, and shear banding (white color results from light scatter). There are also a few visible discrete white lines. Higher-resolution SEM micrographs reveal that these are the bundles of large crazes and/or shear bands. There is a lesser domain of large deformation and more strain localizations (white lines) in form of shear bands within the inner skin and crazes on the rest of the cross section b–b than that observed in a–a. As we move further toward the front of AZ, the number of discrete localizations (crazes and shear bands) decreases, and their thickness reduces. It means the damage density reduces toward the periphery of AZ. It should be noticed that no damage, i.e., no PZ has been observed in front of the crack in the same PB in the test conducted at room temperature, in contrast with the previously described AZ observed at 110°C. Thus, the mechanism of crack growth in this case dramatically changed with temperature. It creates an obstacle for the extrapolation of high temperature accelerated test data to room temperature since the requirement of fracture mechanisms similarity is violated. The reproducibility of fracture mechanism is an important requirement for accelerated testing. In what follows we report experimental examination of the similarity of the mechanism and kinetics of CL growth at various load levels and temperatures in a commercial polyethylene (PE). PE is a semicrystalline thermoplastic from the polyolefin family. PB discussed earlier also belongs to that family.

The aforementioned necking has a direct relation with the failure of engineering plastics. High stress level indeed leads to an early failure; however, it commonly results in a ductile failure associated with highly localized large irreversible deformation. For example, the ductile failure of gas transmission PE pipe illustrated by Figure 5.52 is an example of cold

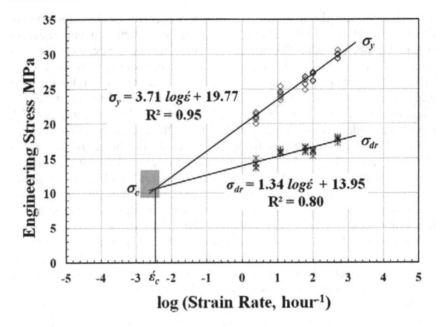

Figure 5.53 Strain rate dependency of yield and drawing stresses in PE.

drawing often called "necking". This type of failure in pipe is referred to as "ballooning". Thus, necking leads to the formation of a thin membrane pushed outward by internal pressure. The necking is well visible on the cross section through the center of the bulge shown in Figure 5.52b. In cold drawing, the original isotropic material is transformed into highly oriented one with corresponding thinning and elongation up to a few hundred percent. The sharp reduction of the pipe wall within the ballooning in comparison with original wall thickness visible on the circumferential cross section illustrates the thinning of the structure associated with orientation. At an intermediate stress level, the failure occurs in a brittle manner, i.e., results from cracking. Thus, one cannot extrapolate high stress and short duration test results into long service life under intermediate stresses due to the change of the failure mechanism from ductile to brittle. Premature failure of plastic components widely reported in forensic literature is mainly attributed to brittle cracking, driven by either mechanical stress or a combination of stresses and chemically aggressive environment. At a low stress level and very long service time, material aging such as chemical degradation rather than mechanical stress becomes the main cause of degradation-driven brittle mode of failure. Such changes of the leading mechanisms of failure with variation of stress level are depicted in Figure 2 of the publication by Chudnovsky et al. (2012). The sketches and photos next to the lines depict the characteristic appearances of the corresponding failure mode. High stress region of the failure has no bearing on extrapolation techniques to predict lifetime but still is utilized in international standards like ISO 9080. Ductile failure is very sensitive to stress variation: A very small change in stress level results in a large variation of time to failure. In contrast with that, the degradation-driven brittle failure weakly depends on stress level. Stress-driven brittle fracture shows an intermediate stress dependency, as can be seen from the slope of the corresponding lines in Figure 2 of the publication [Chudnovsky et al. (2012)]. The change in failure mechanism with variation of stresses depicted in Figure 2 suggests that only a very limited acceleration of failure time can be achieved by varying stress level.

Delayed Necking

Alternative modelling of ductile failure time follows from (a) recognition of ductile failure as delayed necking, (b) a fundamental symmetry between logarithm of strain rate in steady state creep versus stress on one side and logarithm of ductile failure time versus stress on the other side that have been proven more than half a century ago by Hoff [Hoff, N.J., *"Structures and Materials for Finite Lifetime"*, Advances in Aeronautical Sciences", 928, 1959], and (c) the equation expressing yield strength and draw stress–strain rate dependency. The true complexities of the cold drawing in polymers and its departure from classical phase transition are manifested in a strong dependence of cold drawing on strain rate and temperature. We first discuss the strain rate dependency. In ramp tests, both yield and draw stresses decrease with strain rates (Liu et al., 2000; Zhou et al., 1999, 2000, 2005, 2010). The higher value of yield stress compared to draw stress is simply a result of ramp rate (measured as strain rate) that causes overshooting to take place followed by immediate drop in stress level to a draw stress value. Both draw stress and yield stress are not true material parameters but a manifestation of applied strain rate. Thus, one needs to extract a characteristic stress and strain rate which will be a function of testing temperature. Figure 5.53 presents a typical yield stress and draw stress dependence on strain rate. The decrease by two orders of magnitude in engineering strain rate results in reduction of yield stress by about 30%. The draw stresses display lower sensitivity to the engineering strain rate than the yield stress. It reduces only about 10%, when the engineering strain rate decreases by about two orders of magnitude. As it can be seen from Figure 5.53, the difference between the yield and draw stresses is also reduced with the decrease of strain rate. Extrapolating the trend further, one can identify a strain rate, at which the yield stress coincides with the draw stress. The intersection between two lines of σ_y and σ_{dr} dependences on log defines "characteristic stress σ_c" and "characteristic strain rate" (Mostovoy et al., 1967; Zhou et al., 2008, 2010) respectively. At the characteristic strain rate, there is no "overshooting" of

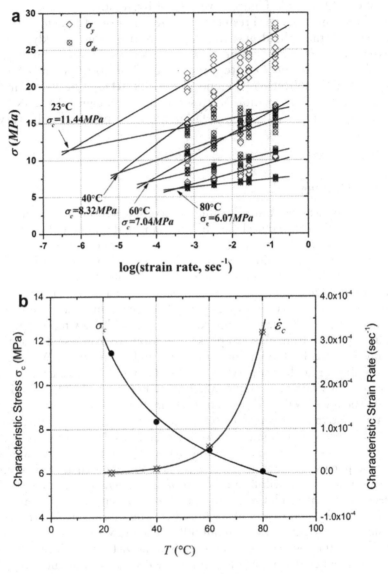

Figure 5.54 (a) Combined effects of temperature and strain rate on yield and drawing stresses; (b) the temperature dependency of the characteristic strain rate and stress.

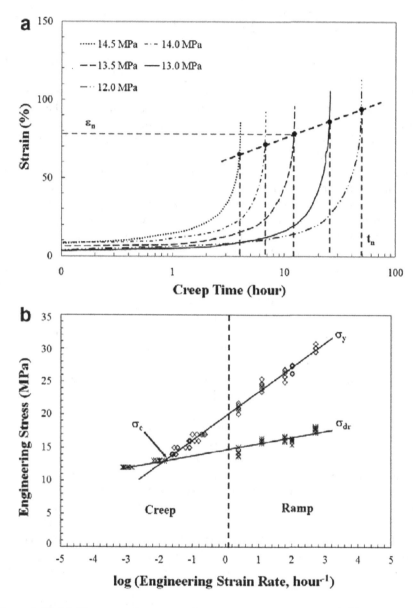

Figure 5.55 (a) Creep strain versus time at various stresses and (b) strain rate effect on yield and drawing stresses.

yield stress over the drawing stress, rather the process of necking progresses at the same characteristic stress in a quasi-equilibrium manner.

The combined effects of temperature and strain rate dependency of yield and draw stresses are depicted in Figure 5.54a. For each temperature, one can identify a characteristic stress value following the same approach as in Figure 5.53. The characteristic stress decreases with temperature, specifically at room temperature it is about twice of that for 80°C. The characteristic strain rate shows the opposite and much stronger temperature dependence: At room temperature, it is about three orders of magnitude lower than that at 80°C. The characteristic stress and strain rate dependence on temperature is presented in Figure 5.54b. It is also noticeable from Figure 5.54 that the characteristic stress and characteristic strain rate are significantly lower than the range of stresses and strain rates used in the ramp tests. Therefore, the extrapolation of data for determining of these characteristics leaves an uncertainty.

Observation of necking under creep conditions allows reducing such uncertainty. Typical data of the creep test at various stress levels is presented in Figure 5.55a. It shows the accelerating growth of creep strain with time until a critical level of strain ε_n is reached at the "time to necking" t_n. At that instance, sudden onset of necking takes place and neck rapidly propagates through the specimen until entire specimen is transformed into an oriented material. It is depicted by the vertical lines, since the length of the specimen increases five to six times over a small fraction of the second. This phenomenon is referred to as "delayed necking" (Chudnovsky et al., 1999; Mostovoy et al., 1967; Zhou et al., 1995). Both the time to necking and strain at necking increase with a decrease of applied stress.

Considering the average creep rate during the steady creep stage and constancy of the applied stress, we can add the stress–strain rate data to the data reported for the ramp test (Figure 5.54). Figure 5.54a displays combined effects of temperature and strain rate on yield and drawing stresses. From this figure, one can extract characteristic strain rate and characteristic stress where yield stress–creep strain rate line and draw stress–creep strain rate converge. Figure 5.54b presents characteristic stress and strain rate as a function temperature leads to two converging curves that meet at a single temperature. Figure 55a displays creep strain versus time at various loading levels. It is evident that as the material continues to creep, it eventually reaches yield and drawing. The yield and drawing stress dependency on strain rate in creep and ramp tests is presented in Figure 5.55b. When applied stress $\sigma_{app} > \sigma_c$, the σ_{app} versus $\log \varepsilon_n$ in creep follows the same straight line as σ_y versus $\log \varepsilon_n$ in ramp test (see Figure 5.55b data above σ_c). However, when applied stress $\sigma_{app} < \sigma_c$, the σ_{app} versus $\log \varepsilon_n$ in creep follows the same straight line as σ_{dr} versus $\log \varepsilon_n$ in ramp test. The two straight lines intersect at characteristic stress, σ_c. Thus, the data shown in Figure 5.55b reinforce the special role of characteristic stress and characteristic strain in cold drawing phenomena as an upper bound for a quasi-equilibrium transition from an isotropic original state to the highly oriented drawn state.

Chapter 6

Solo Brittle Fracture and Statistical Fracture Mechanics

Leonardo Da Vinci (1452–1519) conducted an experiment titled "*Testing the strength of iron wires of various lengths*". A basket was attached at the end of a steel wire and was fed sand through a hopper till it broke. The weight of the sand was then measured to determine tensile strength of the wire (Figure 6.1). Interestingly, he made observations that shorter wires were stronger based on the amount of sand needed to snap the wire. These observations were a bit troubling for early engineers (post-Galileo; 1564–1642) using classical mechanics. According to classical mechanics, the stress (force/area) is same throughout the wire. The failure would be caused when ultimate stress value is exceeded, therefore, the strength should be independent of the length of the wire [Lund, J.R. and Byrne, J.P. (2000)].

It took some time for the early engineers to resolve this conflict. Parsons (1939) notes that the result described, where shorter wires supported a greater weight, this conflicts with the classical theory of mechanics of materials. Classical theory holds that the wire's length should be irrelevant since the stress should be the same along the entire length of such a wire and that the weight of the wire should be negligible compared to the weight in the basket. Parsons' explanation of the recorded result is then that Leonardo Da Vinci, in his notes, mistakenly recorded the experiment and its results [William Parsons, B. (1939)]. This is a very common practice in the scientific community to blame the experimental results when they lack foresight into constructing a hypothesis to challenge or verify.

While Parsons notes that there are frequent errors in Leonardo's notebooks, the particular errors suggested by Parsons, while plausible, seem unlikely. Another alternative explanation rejects the common assumption that the properties of the material are constant throughout and the material is in a continuous medium. However, properties can (and do) vary, and there are flaws in the materials. These inhomogeneity and flaws give rise to higher localized stresses which cause local failures which build up over time to cause ultimate failure. The steel cable manufactured in the times of Da Vinci (or the rope here) would have had many such flaws. Statistically speaking, longer

DOI: 10.1201/9781003359845-7

Figure 6.1 Leonardo da Vinci's experimental set up from his notebook.

cable would have a higher number of (and bigger) flaws. This would lead to more localized failures, and, consequently, it was more likely to fail at lower value of average stress. Thus, probabilistic treatment is necessary to explain the curious observation of Leonardo. Actual material behavior thus obscures the principles of engineering mechanics [Lund, J.R. and Byrne, J.P. (2000)]. In later years, the effects of increasing size (including diameter and length) on increasing the probability of imperfections and reductions in strength were developed statistically by Weibull (1939). This behavior was demonstrated for high phosphorus steel in experiments by Davidenkov et al. (1947), somewhat similar to those by Leonardo Da Vinci and also showing failure stress decreasing for longer samples. Only domain Weibull Statistics do not hold in nearly defect-free systems like nanotube/graphene. The Weibull distribution is related to a number of other probability distributions; in particular, it interpolates between the exponential distribution and the Rayleigh distribution.

In vain Classical Physicists tried to eliminate chance from foundation of Physics. They believed everything is deterministic and they can predict based on Newton's laws. They can define the initial condition with high precision. But due to uncertainty principle in quantum mechanics we know now that is a false assumption.

Max Born—*My Life and Views*
Charles Scribner's Sons
New York, 1965

Similarly, conventional Fracture mechanics has tried to ignore the role of chance in fracture phenomena. Leonardo's work was the earliest indication that even strength σ_c is not always a single value for brittle linear elastic materials as expected from classical theory of strength due to the presence of defects. One would expect that K_c or G_c introduced by Fracture mechanics will address this inconsistency by taking into consideration of the crack size. Indeed, many materials' toughness could be characterized by a single value of K_c and G_c for a specific fracture mode, namely for polymers from epoxy family, large varieties of inorganic glass and ceramics.

CL theory introduced earlier described the propagation of a crack coupled with an evolution of the damage zone (cooperative brittle fracture) by a set of deterministic equations. Another extreme mode of failure is a perfectly brittle solo fracture when a crack propagation is controlled by a preexisting field of defects and does not cause noticeable changes to this field. In this case the random location and orientation of the individual microdefects result in an irregular, stochastic crack trajectory; scatter of the main fracture parameters; and a scale effect. Statistical Fracture mechanics was first introduced by A. Chudnovsky at the A.I. Lurie's seminar at the Leningrad Polytechnic Institute in the fall of 1969. After a long and heated discussion, Professor A.I. Lurie suggested presenting the work as a thesis for a Doctor of Science in Physics and Mathematics. In 1969, Alexander Chudnovsky formally proposed statistical Fracture mechanics introducing a virtual set of fracture trajectories and proposed a thought experiment to observe such a phenomenon [Chudnovsky, A. (1973), Chudnovsky, A. (ed. L. Kachanov) (1973)]. A few years later, A. Chudnovsky moved to the United States, where he continued this work with his graduate students at Case Western Reserve University.

The first set of experiments were conducted by the PhD graduate student M.A. Mull under the supervision of A. Chudnovsky and A. Moet at the Case Western University in Cleveland, Ohio. The results were submitted to the Philosophical Magazine 20 years after the experiment was first thought of. Pioneering work by Mull and Chudnovsky brought to our attention an order of magnitude scatter in fracture toughness G_{1c} in a micro-heterogeneous composite. The material used in this case was a polyester composite reinforced with glass beads and Kevlar (poly-paraphenylene terephthalamide) spun fiber [Mull et al. (1987)]. This type of composites are extremely rigid and fairly linear elastic in their stress–strain response. Fatigue cracks in this composite propagate with no detectable damage along random trajectories, reflecting severe heterogeneities in the composite. This study was conducted on polyester reinforced with 9 vol% 1/8 inch chopped short Kevlar-29 fibers and 14 vol% inorganic glass bead fillers. The analysis part of this work was based on previous developments by Chudnovsky, A. and Perdikaris, P. (1983).

Edge-notched specimens (prepared from 12-inch compression molded plaques of aforementioned material, 1.5 mm thick; a 60 V-notch, 1.5 mm deep, was machined at the center of one edge of the specimen to initiate crack propagation) were used for the aforementioned fatigue crack propagation tests conducted at room temperature. Sinusoidal loading was used with a load ratio of 0.02 and a frequency of 0.5 Hz. Fatigue crack was observed to propagate with no detectable damage, controlled by local fluctuation of material structure. This type of fracture was classified as "single path" failure or solo brittle fracture. Crack trajectories in the composite studied were observed to deviate from the usual rectilinear path, perpendicular to the direction of the applied load, as much as by 30° in either direction (Figure 6.2). Set of crack trajectories from 25 fracture toughness test is shown in the figure. The solid bullet on each trajectory identifies the onset of catastrophic failure

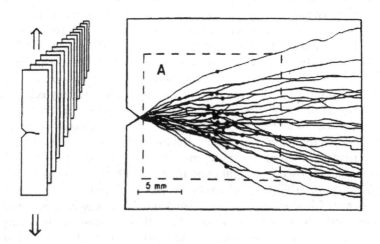

Figure 6.2 An ensemble of macroscopically identical specimens cracked under identical fatigue load and the superposition of the profilograms of the crack traced from each individual specimen (only one trajectory takes place in each specimen).

in each test. Each trajectory is unique to a given sample; no two trajectories are ever identical because of random nature of the strength field. Although the samples are macroscopically identical, the microscopic nature of each sample is unique. It is presumed that a propagating crack follows the weakest path through the material. Fracture toughness is a measure of the material's resistance to crack propagation. For a random crack path, fracture toughness is a function of the crack trajectory. Thus, fracture toughness becomes a random variable which can be characterized by its moments, mathematical expectation, variance, etc. As the material becomes more homogeneous, the scatter in fracture toughness decreases; that is, the variance tends to vanish, and the mean value (mathematical expectation) becomes the only fracture toughness parameter.

For ideally brittle homogeneous material, Griffith's criteria encompass a necessary and sufficient condition for instability. By choosing the sample geometry for this study, sufficient condition is met automatically (see Chapter V). Thus, $G_c = 2\gamma$ (creation of two surfaces) meets both necessary and sufficient conditions for homogeneous material. For a heterogenous material, however, Griffith's criteria met at a single point on the trajectory may not be met at another point, because of fluctuations in the random strength field. Therefore, for a failure to occur, Griffith's criteria should be satisfied at each point of the crack trajectory. Thus, this criterion is global in the sense that the entire trajectory of the crack contributes to the instability condition, whereas original Griffith's condition is a local criterion. Trajectories from a set of 25 (identically prepared and loaded) samples have been superimposed upon the same set of coordinate axes. The propagating crack possesses a deterministic forward movement owing to the stress field applied as well as a lateral movement owing to fluctuations of the strength field $\gamma(x,y)$. The enhanced fracture toughness of reinforced composites results from their ability to deviate the crack from its most efficient path, that is, from the trajectories, and it can be taken as a characteristic of toughness.

Since the crack trajectories are skewed, the calculation of G differs from that for a rectilinear path. Thus, a mixed-mode solution of the elastic energy release rate G was considered $\left(G = K_I^2 + K_{II}^2 \,/\, E\right)$. In each trajectory, the crack propagated till the transition from stable SCG to avalanche-like propagation is observed. Depending on loading history, this transition occurred at a certain combination of stress and crack length. These are traditionally identified as critical stress σ_c and critical crack length l_c. Using σ_c and l_c, G_{1c} was calculated for each individual trajectory. A statistical approach was introduced to evaluate the cumulative probability of the critical energy release rate, G_{1c}, as a function of fracture stress, critical crack length, and the angles of crack trajectory inclinations which were random variables extracted from the test results. It was observed that the assemblage of crack trajectories resembles features of Brownian particle motion. Considering crack propagation as a diffusive process, a "crack diffusion coefficient" was introduced which reflects the ability of the composite to deviate crack trajectory from the energetically favorable path and thus linked material toughness to its structure. Assuming that crack propagates through the "weakest" points, the strength field γ was assumed to follow Weibull distribution. The field was characterized by scale and shape parameters with a radius of correlation. In this extreme type of composite failure, no detectable damage associated with the crack was observable from the side view. (In probability theory and statistics, the Weibull distribution is a continuous probability distribution. It is named after Swedish mathematician Waloddi Weibull, who described it in detail in 1951, although it was first identified by Maurice René Fréchet and first applied by Rosin and Rammler (1933) to describe a particle size distribution.)

The original proposition by Chudnovsky gradually evolved into the formulation of Statistical Fracture Mechanics (SFM). The probability of a brittle crack formation in an elastic solid with fluctuating strength was considered by Chudnovsky and Kunin (1987, 1992). In their paper, they distinguished two extreme cases of the influence of defects on the fracture process, and modelling of these two processes requires different formalisms.

Case I: The intensity of damage formed as a response to the stress concentration at the tip of a propagating crack is much greater than the intensity of the preexisting damage. The crack propagation is then inseparable from the evolution of the damage accompanying the crack. Crack layer theory discussed earlier addresses this case.

Case II: The other case is the case of solo crack growth where crack propagates through a field of preexisting defects, causing negligible changes to the field (Figure 6.3). The fluctuation of the microdefect field is directly reflected in stochastic features of fracture surfaces and leads to the scatter of the experimentally observed fracture parameters such as critical crack length and critical load. A probabilistic approach seems most adequate in this case.

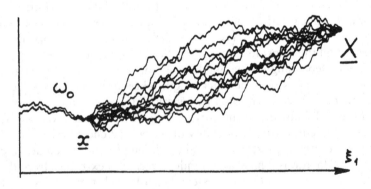

Figure 6.3 Probability of crack growth from x to X: P{x,X}.

The apparent randomness of brittle fracture is closely associated with the distribution of defects on various scales within a solid. The presence of microdefects is modeled by a random field of specific fracture energy γ following the framework of SFM. A brief summary of SFM is presented. SFM is the only model which explicitly incorporates the fractographic information, e.g., fractal characterization of fracture surfaces in the probabilistic description of brittle fracture. At the same time, the model has limitations in engineering applications, mainly due to its mathematical complexity. In this paper, the Monte Carlo technique is employed to overcome these limitations. It allows one to combine the physical insight and modeling of the fracture mechanisms in SFM with the flexibility of the Monte Carlo method. Probability distributions of the fracture parameters such as critical load, critical crack length, and fracture toughness are simulated and compared with experimental observations.

Dependency of the conventional measure of fracture toughness on roughness of crack profiles, specimen, and grain size, as well as load level is discussed. The ambiguity of the concept of fracture toughness in a probabilistic setting is addressed. Statistical analysis of the observed crack trajectories allows one to formulate reasonable assumptions about the nature of the set Ω of all possible crack trajectories for each specimen under given test condition. If we assume that only one crack is formed beginning at the notch and extending to or beyond the depth X, probability can be written as $P(X) = \int_{\Omega} P\{X \, / \, \omega\} \cdot d\mu\{\omega\}$. Following this logic, a set $\Omega_{x,X}$ of all possible crack trajectories (ω) reflecting the fluctuation of the strength field is introduced. The probability $P(\bar{x}, \bar{X})$ is that crack propagation depth from $\bar{x}$ to $\bar{X}$ is expressed as a functional integral over $\Omega_{x,X}$ of a conditional probability of the same event taking place along a particular path.

$$P(\bar{x}, \bar{X}) = \int_{\Omega_{x,X}} P\{\bar{x}, \bar{X} \, / \, \omega(x)\} \cdot d\mu\{\Omega_{x,X}\} \qquad (6.1)$$

The evaluation of the integral leads to solving a diffusion-type equation by modelling crack trajectories in Brownian paths ($y = \omega(x)$, $0 \le x \le X$, $\omega(0) = 0$) ($\Omega_x = \{\omega\}$ will denote the space) and choose $d\mu\{\omega\}$ to be a Wiener measure [Gefland et al. (1956)]. In mathematics, the Wiener process is a real-valued continuous-time stochastic process named in the honor of American mathematician Norbert Wiener for his investigations on the mathematical properties of the one-dimensional Brownian motion. It is often also called Brownian motion due to its historical connection with the physical process of the same name originally observed by Scottish botanist Robert Brown.

The Wiener process plays an important role in both pure and applied mathematics. It is a key process in terms of which more complicated stochastic processes can be described. The Wiener process has applications throughout the mathematical sciences. In physics, it is used to study Brownian motion, the diffusion of minute particles suspended in fluid, and other types of diffusion. It also forms the basis for the rigorous path integral formulation of quantum mechanics. It is also prominent in the mathematical theory of finance. Thus, the probabilistic measure in this case can be expressed as:

$$d\mu^D\left(\Omega_{x,X}\right) = Cexp\left[\int_{x_1}^{X_1}\left(\omega'(x)\right)^2 \frac{dx}{2D}\right]\prod_{x=x_1}^{X_1} d\omega(x)\tag{6.2}$$

A new characteristic of fracture process, "crack diffusion coefficient D", is introduced in this work. D reflects the tendency of crack trajectories to deviate from the X axis and is experimentally measurable. From the thermodynamics consideration, crack trajectory will choose the path where Griffith criterion is met. In a homogeneous material like inorganic glass, Griffith demonstrated his theory successfully using a single value of surface energy γ. However, in a heterogeneous material, γ becomes a random field (γ-field) and global instability criterion should be written as an integral requirement that G_1 or $J_1 > 2\gamma$ is met everywhere along a fracture path. Thus, a generalized Griffith condition is introduced and expressed in Eq. (6.2) as conditional crack propagator $P\{\bar{x},\bar{X}/\omega(x)\}$. Figure 6.4 shows a crack trajectory from a point $\bar{x}$ to $\bar{X}$ with an incremental growth of Δx_k. This leads to Eq. (6.3).

$$P\{\bar{x},\bar{X}/\omega(x)\} = \lim_{n\to\infty}\prod_{n=1}^{N} P\{G_1(x,\omega) > 2\gamma(x)\}\tag{6.3}$$

Following Weibull statistics, the probability of crack arrest $U(x)$ can be expressed as:

$$U(x) = exp\left(-\left[\frac{G_1(x,\omega) - 2\gamma_{min}}{2\gamma_0}\right]^\alpha\right)\tag{6.4}$$

where α, γ_0, and γ_{min} are Weibull distribution parameters.

The parameter D that emerged from this development reflects the tendency of crack trajectories to deviate from the rectilinear path ahead of starting notch for a single-edge notch geometry. As explained earlier, it is also the most efficient path corresponding to the maximum energy-release rate. Thus, D can be taken as a characteristic of toughness for such solo crack instead of G_c. Larger the D, higher the toughness of a material displaying solo crack growth of varying trajectories in specimen to specimen. It is worth mentioning that fracture surfaces in brittle materials are never a smooth surface but contain irregularities. These irregularities stem from the irregularities in crack trajectories, source of which lies in the random γ-field instead of a single γ explained earlier.

The problem of adequate modeling of actual crack trajectories has been addressed by Kunin and Gorelik (1991). Because of the apparent discrepancy

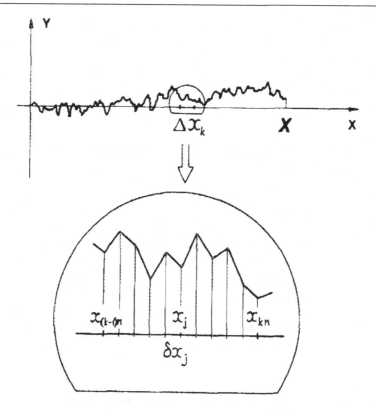

Figure 6.4 A crack trajectory from a point $\bar{x}$ to $\bar{X}$ is shown with incremental growth of Δx_k.

between the observed fracture paths and Brownian trajectories, it was suggested to use partially smoothed Wiener paths. Partial smoothing can be achieved by means of fractional integration of Wiener process. Let w (x) be a conventional Wiener process. The fractional integral of w (x) to the power λ is defined as [Lavoie et al. (1976)]:

$$w_\lambda(x) = \frac{1}{\Gamma(\lambda)} \int_0^x w(\xi)(x-\xi)^{\lambda-1} \, d\xi \tag{6.5}$$

with $0 < \lambda < 1$, where $\lambda = 1$ corresponds to the conventional Cauchy integral, and $\lambda < 1$ determines different degrees of smoothing. Parameter λ is simply related to the fractal dimension d of $w_\lambda(x)$: $d = 1.5 - \lambda$ (for $\lambda \sim 0.5$) [see Appendix II]. Today, these integrals can be solved numerically, but authors resorted to Monte Carlo approach discussed later in the chapter. In order to predict the observation reported in Figure 6.2 by Mull et al., Gorelik et al.

proposed a modified Wiener process (Brownian process is a one-dimensional Wiener process) for crack trajectories instead of Brownian trajectories as proposed by Chudnovsky et al. (1987a, b). This approach led to a fractional integration of the Wiener function to a power λ where $0 \leq \lambda \leq 1$. Thus, Wiener process with lower value of λ will lead to a rougher fracture surface represented by higher fractal dimension. It was also shown by the researchers that the variance of the Wiener process is related to the crack diffusion coefficient D introduced earlier. This makes perfect sense from the earlier work by Boris et al. where it is claimed that larger D represents greater toughness. A computer simulation of 27 realizations of Wiener process with $\lambda = 0.4$ (d = 1.1) is shown in Figure 6.5. The simulated sample set of trajectories closely resembles the set of experimental trajectories in Figure 6.2.

One way to deal with the mathematical complexity of SFM in the form published by these authors limits the engineering application of the theory [Chudnovsky, A. and Kunin, B., November 1987, 1992)]. However, SFM is the only model which explicitly incorporates fractographic information, e.g., fractal characterization of fracture surfaces in the probabilistic description of the brittle fracture. Employment of the Monte Carlo simulation technique opens the door for the application of SFM to a wide variety of practical problems. The new method is examined by comparison of computer simulation of the statistics of crack instability with the experimental results reported by

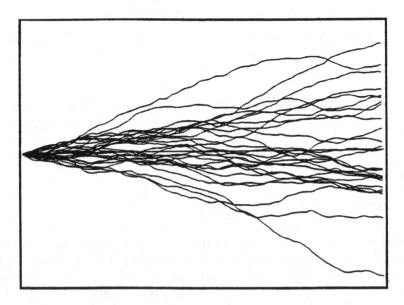

Figure 6.5 A sample set of realizations of the Wiener process for $\lambda = 0.4$, resulting from fractional integration.

Mull et al. (1987) [Chudnovsky, A. and Gorelik, M., January 1994; Chudnovsky et al., November 1997]. This comparison is presented in Figure 6.5 applying the diffusion approximation.

If one observes carefully, it is noticeable that there is a certain similarity between the turbulence in a fluid flow and the fracture in the solid state since the characteristic features of critical phenomenon such as hierarchy of interacting defects, stochastic realizations, and a distinct scale effect are present in both processes. An irregular stochastic appearance of the fracture surfaces is well documented and widely discussed phenomenon. One may suggest two potential sources of such appearance: (a) Intrinsic instability of crack growth process as proposed by some Fracture mechanics researchers and/or (b) an impact of a randomly distributed array of the preexisting material inhomogeneities (defects). According to numerical analysis performed by Gorelik et al., crack trajectories are also stable with respect to a kink-type perturbation [Chudnovsky and Gorelik (1994)]. Thus, within the assumptions of linear Fracture mechanics, the erraticism of brittle fracture surfaces cannot be attributed to the instability of the crack trajectory. Therefore, it can be concluded that the main source of stochasticity in brittle fracture is a random distribution of the material inhomogeneities (microdefects). The presence of microdefects within the Continuum Mechanics context can be modeled by a random field of the specific fracture energy (γ-field). SFM is the only model which explicitly incorporates fractographic information, e.g., fractal characterization of fracture surfaces in the probabilistic description of brittle fracture.

SFM is based on a few natural assumptions regarding the process of brittle fracture as proposed by Chudnovsky and Kunin (1992): (a) the crack path is random, i.e., the crack randomly selects a path from a set of virtual paths; (b) crack advance along a particular path consists of a sequence of local failures in front of the crack tip, controlled by the Griffith criterion G_{1c} or $J_{1c} = 2\gamma$; and (c) the local failures are random events due to the random field $\gamma(x)$. One of the main building blocks of SFM is a Crack Propagator (CP) which is defined by the authors as a probability that the crack extends from crack tip to another point along a certain trajectory that the condition G_{1c} or $J_{1c} = 2\gamma$ is met at every point of the crack trajectory. Assuming that a crack "searches" for the easiest direction at each point of its trajectory, one can conclude that the distribution of γ along the crack trajectory is sampled from the minimal values of the γ field. In this work, they employed Weibull distribution (one of the most popular distributions of extremes) for the values of γ at every point of the crack trajectory.

One big advantage of the numerical simulation is to be able to predict outcomes that cannot be experimentally determined from a limited set of specimens. For each simulated crack trajectory, the value of the critical crack length is found according to the criterion stated earlier. Set of the crack instability points represents a statistical output of the model in this case. Both one-dimensional and two-dimensional probability densities can be readily obtained. Based on the comparison of the first two moments of the experimental and simulated critical crack depths, the optimal parameters of Weibull distribution were found. Based on these parameters, the simulation was performed for 3,000 trajectories, and the output was used to build a two-dimensional distribution of critical crack tip locations. Figure 6.6 displays contour lines with equal levels of probability density for the crack tip position at the point of crack instability. Some of the experimental points (crack tip position at crack instability) are thrown in Figure 6.6 to demonstrate that most of the actual data points are toward the center of high probability, and only a few data points are toward the boundary which is quite sensible. The output of the simulation procedure, described before, is a sample set. Thus, the probability distribution of any critical fracture parameter can be readily reconstructed from this computational work. These parameters include the scale effect, the scatter of fracture toughness, critical energy release rate (ERR) dependency on the loading conditions, and the difference between the ERR values at the crack arrest and initiation.

Now that the model is validated experimentally using Mull et al.'s study, a large number of simulated tests can be done to relate fracture toughness

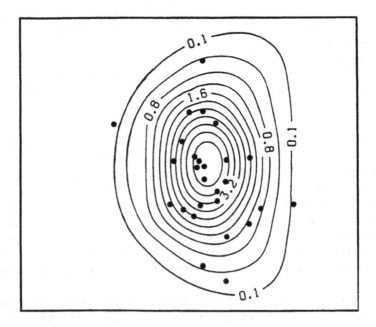

Figure 6.6 Contour lines represent the equal levels of probability density for the crack tip position at the instance of crack instability (Monte Carlo simulation). Dots indicate the experimental points.

with the roughness of crack trajectories. As discussed, earlier fractal dimension can be used as a measure of roughness with d = 1.5 being Brownian crack trajectory, while d = 1 represents smooth trajectory. With an increase of fractal dimension, similar trends are observed for the mean values of the critical G_{1c} and the critical crack depth X_c. Both increase nonlinearly with d. Figure 6.7 describes fractional Brownian trajectories with different fractal dimensions. Value of < G_{1c} > varies by as much as 70%. The computational algorithm had also been used to study the load level effect on the critical ERR such as mean value of < G_{1c} > and standard deviation. Both increase with the increase of applied load σ^{∞}.

One of the typical features of brittle fracture, associated with intrinsic stochasticity of the process, is a scale effect. It is manifested in dependency of fracture toughness parameters on the macroscopic size and shape of the specimen. Linear elasticity is commonly accepted for modeling of brittle fracture. Conventional elastic models do not have an internal scale and, therefore, cannot directly account for material heterogeneities on the microscale. In SFM, those are accounted for by the random field of specific fracture energy (SFE) with the correlation distance. Thus, in the frame of SFM, a change in the correlation distance of the γ-field in the specimen with fixed dimensions is equivalent to the corresponding change in size of the specimen under fixed correlation distance. Specifically, n-times increase in correlation distance is equivalent to n-times decrease in specimen's dimensions. It should be noted

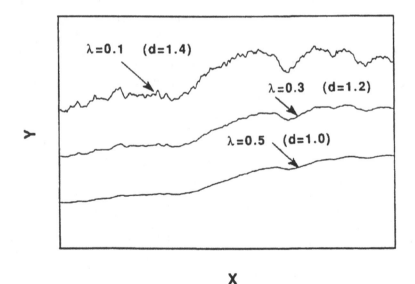

Figure 6.7 Three realization of the fractional Wiener process with λ = 0.1, 0.3, and 0.5. Corresponding fractal dimension is indicated in parenthesis.

that a change of correlation distance may be interpreted as a replacement of the specimen's material by another one with the same elastic properties, but different microstructure, while the specimen size and shape are fixed.

One of the typical features of brittle fracture, associated with intrinsic stochasticity of the process, is a scale effect. The scale effect in our model was studied by changing the relative correlation distance. It is manifested in the dependency of fracture toughness parameters (K_{1c}, G_{1c}) on the macroscopic size and shape of the specimen. Linear elasticity is commonly accepted for the modeling of brittle fracture. Conventional elastic models do not have an internal scale and, therefore, cannot directly account for material heterogeneities on the microscale. In SFM, those are accounted for by the random field of SFE (γ) with the correlation distance r_0. Thus, in the frame of SFM, a change in the correlation distance r_0 of the γ-field in the specimen with fixed dimensions is equivalent to the corresponding change in size of the specimen under fixed r_0. Specifically, n-times increase in r_0 is equivalent to n-times decrease in specimen's dimensions. It should be noted that change of r_0 may be interpreted as a replacement of the specimen's material by another one with the same elastic properties, but different microstructure, while the specimen size and shape are fixed. The scale effect in our model was studied by changing the relative correlation distance r_0 / W. Hundred statistical trials were performed for correlation distances of 0.003 and 0.009. For each value of correlation distance, 100 crack trajectories were simulated, and local values of the surface energy γ and G_1 were generated for every trajectory. Comparing G_1 with 2γ pointwise along a particular path, a set of critical crack depths and corresponding critical G_1 values were obtained. Statistical output in the form of probability densities (approximated by Weibull distribution) of critical crack depth and critical G_1 was made, and the distribution of correlation for smaller correlation distance is shifted to the left and has a larger relative scatter. Distribution of G_1 for smaller correlation distance is also shifted to the left, but the difference in relative scatter is less pronounced.

Brittle fracture is one of the best illustrations for the role of a chance in physical phenomena. The main source of stochasticity in fracture is the population of the microdefects resulting from particular technological process. Statistical Fracture mechanics bridges the conventional Fracture mechanics with the weakest link theories. The elements of Fracture mechanics in SFM reflect the fracture localization phenomena. The concepts of the weakest link are employed in SFM to account for the mechanism controlled by chance. Computer-simulated experiment allows to study different effects typical for the brittle fracture, such as scale effect and load level effect. The method can be applied to a broad range of loading conditions, specimen geometries, and types of fracture surfaces. One of the typical features of brittle fracture, associated with intrinsic stochasticity of the process, is a scale effect. It is manifested in the dependency of fracture toughness parameters (G_{1c}) on the macroscopic size and shape of the specimen. The scale effect can be addressed experimentally as well and will be discussed in the next section.

Scale Effect in Toughness

An important question to ask is how to extrapolate small scale laboratory test results to a large size structure (size scaling)? In order to address that, let us take a quick peek at how scientists addressed relationship between bone size and weight of any creature (Figure 6.8). Consider the weight of the animal, $W = \rho V \sim m^3$, and bone cross-sectional area $A \sim m^2$, with load-bearing capacity $S = A / W \sim m^{-1}$. In that case, load-bearing capacity per unit weight should be $S \sim W^{-\frac{1}{3}}$ (Dialogues, *Two New Sciences*, Galileo 1638). Following this logic, load-bearing capacity, S_{lb} of a ladybug with $W \sim 0.5$ g is more than 215 times load-bearing capacity S_{el} of an elephant with $W \sim 5$ tons (Figure 6.9). On similar vein we can ask ourselves "Is there a scale effect in toughness?".

Figure 6.8 Elephant represents a large bone size creature with large weight.

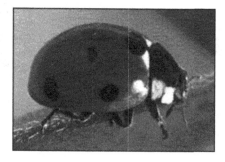

Figure 6.9 A ladybug representing a small creature.

An approach to address this in materials that will predominantly undergo solo fracture is to design experiments to highlight a scale effect. To demonstrate this, Issa et al. picked concrete as the material of choice. All specimens were cast using a specific type of cement, and the mix was made to have certain minimum strength [Issa et al. (1993)]. The water–cement ratio was kept constant, and several different maximum aggregate sizes of crushed limestone were used in this study. Compact tension (CT) specimens were used with a prescribed initial notch (a_0)/depth (d_0) ratio, and a similarity approach was utilized for setting up the specimen thickness/aggregate size ratio. The experiments were done on the CT specimens maintaining the ratio of specimen thickness to size (W) approximately constant (Figure 6.10). Figure 6.10 displays scaling of specimen size. Displacement controlled tests were done in a servo-hydraulic testing machine, and the measurements of crack mouth opening displacements (CMOD) were recorded as a function of load (Figure 6.11). By continuously monitoring the load–CMOD curve on the screen, the specimens were monotonically loaded up to a maximum load corresponding to the initiation of rapid crack growth (crack length = a) and then unloaded. Subsequently, the specimen was reloaded and so on till the last cycle where the specimen was loaded to final failure.

Figure 6.10 Scaling of specimen size.

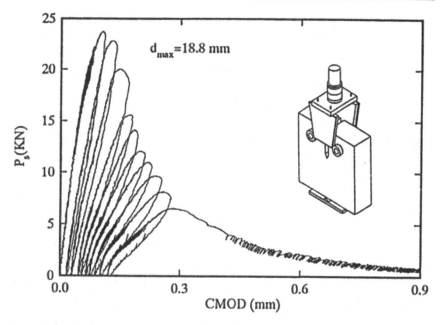

Figure 6.11 A typical loading, unloading, and reloading versus crack mouth opening displacement curve.

Acoustic emission (AE) technique was used to mark the initiation of rapid crack growth. Random crack trajectories were recorded in identical specimens demonstrating the statistical nature of fracture progressing through a field of preexisting defects. Fracture in concrete is complicated by its heterogeneity which is well manifested in the irregularities on the fracture surface topography. To characterize the fracture surface geometry in the present study, Issa, Hammad, and Chudnovsky introduced a new nondestructive technique for fractal dimension D evaluation for concrete and to correlate with fracture toughness [Issa et al. (1993)]. Details of this technique are elaborated in the cited paper. It was used as an alternative to the well-known slit-island technique proposed by Mandelbrot et al. that requires destructive procedures [Mandelbrot et al. (1984)]. Results of the analysis indicate that the concept of fractal geometry provides a useful tool in the fracture surface characterization that can be correlated with the fracture toughness. It was found to be a semilogarithmic relationship between G_{1c} (fracture toughness) and excess of fractal dimension ΔD above 2. It is well known that fractal dimension of 2 represents a smooth Euclidean surface.

$$G_{1c} = G_{1c}^0 \exp(K \cdot \Delta D) \tag{6.6}$$

G_{1c}^0 corresponds to the ideal, plane fracture surface. Further investigation is needed to examine whether the relationship Eq. (6.6) holds for various specimen geometries, and G_{1c}^0 is a true material parameter. In other words, this supports the existence of a correlation between fracture toughness and fractal dimension (roughness).

G_{1c} also was found to increase with an increase of the maximum aggregate size, d_{max}, used to prepare the concrete. It also involved changing the $\dfrac{a}{W}$ ratio from 0.2 to 0.6 in increment of 0.1 to evaluate G_{1c}. Combining all the results, a functional relationship between critical G_{1c} and original G_{1c}^0 associated with the starter notch was expressed as:

$$G_{1c}\left(a, \frac{a}{W}, W\right) = G_{1c}^0\left(a_0, \frac{a_0}{W_0}\right) * \left(\frac{W}{W_0}\right)^\alpha \quad (6.7)$$

where the parameter α turned to be 0.64. Thus, for concrete, a scaling relationship with specimen size W was established, i.e., G_{1c} increases with size as approximately $W^{0.64}$ since all other parameters in the aforementioned relationship are constants. Coincidentally, about 400 years ago, Galileo proposed that load-bearing capacity per unit weight should be $S \sim W^{-0.33}$. This proposal was never confirmed.

Summary (Bridge between Statistical Fracture Mechanics and Crack Layer)

The crack propagator of the SFM can be visualized as a probability cloud hanging above a domain of concentration of the crack-tip stress field ahead of the crack (analogous to a probability cloud of a particle representing a distribution through a region of space of the probability of detecting a given particle) [Chudnovsky (1973), Chudnovsky and Kunin (1987, 1992), Chudnovsky, A. and Gorelik, M. (1994)]. The probability cloud [Chudnovsky, A. and Kunin, B. (1987, 1992)] is materialized as a damage zone which is also called the Process Zone (PZ). It consists of microscopic and submicroscopic defects formed due to the high stress induced by the crack. A system of crack and the PZ is named Crack layer (CL). CL evolution is discussed in earlier publications [Chudnovsky, A., a Review paper (2014), Kunin, B. et al. (1973)].

References

Adams, A.W., *A Textbook of Physical Chemistry*, Academic Press, London, New York, San Francisco, pp. 191–194 (1979).

Anderson, T.L., *Fracture Mechanics: Fundamentals and Applications*, CRC Press, Boca Raton, Ann Arbor, Boston (1991).

Ashby, M.F. and Tomkins, B., *Proceedings, ICM-3 (1979)* (F. Smith and G.A. Miller, eds.), Pergamon, Oxford, Vol. 1, p. 47 (1980).

Atkinson, C., "The Interaction between a Dislocation and a Crack", *International Journal of Fracture*, Vol. 2, pp. 567–575 (1966).

Bakar, M., Haddaoui, N., Chudnovsky, A. and Moet, A., "The Effect of Loading History on the Fracture Toughness of Polymers", *Proceedings SPE-ANTEC*, pp. 544–547, Chicago (May 1983).

Ballarini, R. and Denda, M., "The Interaction between a Crack and a Dislocation Dipole", *International Journal of Fracture*, Vol. 37, pp. 61–71 (1988).

Barenblatt, G.I., "The Mathematical Theory of Equilibrium Cracks in Brittle Fracture", *Advances in Applied Mechanics*, Vol. 7, pp. 55–129 (1962).

Ben Ouezdou, M. and Chudnovsky, A., "Semi-Empirical Crack Tip Analysis", *International Journal of Fracture*, Vol. 37, pp. 3–11 (1988).

Benham, P.P. and Warnock, F.V., *Mechanics of Solids and Structures*, Pitman, London (1973).

Bennett, A. and Mindlin, H., "Metallurgical Aspects of the Failure of the Point Pleasant Bridge", *Journal of Testing and Evaluation*, Vol. 1, No. 2, p. 152 (1973).

Berdichevsky, V. and Khanh, L., "On the Microcrack Nucleation in Brittle Solids", *International Journal of Fracture*, Vol. 133, pp. 47–54 (2005).

Born, M., "Thermodynamics of Crystals and Melting", *The Journal of Chemical Physics*, Vol. 7, No. 8 (1939).

Born, M., "On the Stability of Crystal Lattices", Part 1, *Proceedings of the Cambridge Philological Society*, Vol. 36, pp. 160–172 (1940).

Botsis, J., "Crack and Damage Propagation in Polystyrene Under Fatigue Loading", *Polymer*, Vol. 29, No. 3, pp. 457–462 (March 1988).

Botsis, J., "Damage Analysis of a Crack Layer", *Journal of Materials Science*, Vol. 24, No. 6, pp. 2018–2024 (June 1989).

Botsis, J., Chudnovsky, A. and Moet, A., "The Effect of Damage Dissemination on Fracture Propagation", Proceedings of the 6th International Conference on Fracture (ICF6), New Delhi, India, pp. 2595–2602 (December 4–10, 1984).

Botsis, J., Chudnovsky, A. and Moet, A., "Fatigue Crack Layer Propagation in Polystyrene—Part I Experimental Observations", *International Journal of Fracture*, Vol. 33, No. 4, pp. 263–276 (1987).

Botsis, J., Chudnovsky, A. and Moet, A., "Fatigue Crack Layer Propagation in Polystyrene—Part I Experimental Observations", *International Journal of Fracture*, Vol. 33, No. 4, pp. 277–284 (1987/4).

Botsis, J., Moet, A. and Chudnovsky, A., "Microstructure of Crack Layer in Polystyrene Under Cyclic Loads", *Plastics Engineering*, Vol. 39, No. 3, p. 52 (January 1983).

Broek, D., *Elementary Engineering Fracture Mechanics*, Noordhoff, Leiden (1974).

Brown, W.K., Krapp, R.R. and Grady, D.E., "Fragmentation of the Universe", *Astrophysics, and Space Science* (ISSN 0004–640X), Vol. 94, No. 2, pp. 401–412 (August 1983).

Buchel, A. and Sethna, J.P., "Elastic Theory Has Zero Radius of Convergence", *Physical Review Letters*, Vol. 77, pp. 1520–1523 (1996).

Carlsson, J. and Isaksson, P., "Crack Dynamics and Crack Tip Shielding in a Material Containing Pores Analysed by a Phase Field Method", *Engineering Fracture Mechanics*, Vol. 206, pp. 526–540 (February 1, 2019).

Chan, M.K. and Williams, J.G., "Slow Stable Crack Growth in High Density Polyethylene", *Polymer*, Vol. 24, p. 234 (1983).

Chen, C., Pan, E. and Amadei, B., "Determination of Deformability and Tensile Strength of Anisotropic Rock Using Brazilian Tests", *International Journal of Rock Mechanics and Mining Sciences*, Vol. 35, pp. 43–61 (1998).

Choi, B.H., Paradkar, R., Cham, P.M., Michie, W., Zhou, Z. and Chudnovsky, A., "SEM and FTIR Analysis of PE Pipe Fracture in Accelerated Test Conditions, Experimental and Theoretical Investigation of Stress Corrosion Crack Growth in Polyethylene Pipes", *Polymer Degradation and Stability*, Vol. 94, p. 859 (2009).

Choi, B.H., Zhou, W. and Chudnovsky, A., "Fracture Initiation in Polybutylene Tubing in Potable Water Applications", 61st Annual Technical Conference, SPE ANTEC, Nashville, Tennessee (2003).

Choi, B.H., Zhou, Z., Chudnovsky, A. and Sehanobish, K., "Stress Corrosion Cracking in Plastic Pipes: Observation and Modelling", *International Journal of Fracture*, Vol. 145, p. 81 (2007).

Choi, B.H., Zhou, Z., Chudnovsky, A., Stivala, S., Sehanobish, K. and Bosnyak, C.P., "Fracture Initiation Associated with Chemical Degradation: Observation and Modeling", *International Journal of Solids and Structures*, Vol. 42, p. 681 (2005).

Chudnovsky, A., In Collection of Scientific Works on Elasticity and Plasticity, 9, Article, *Foundations of Statistical Fracture Mechanics* (L.M. Kachanov, ed.), St. Petersburg State University, St. Petersburg, Leningrad, pp. 2–39 (1973).

Chudnovsky, A., "On Fracture of Solids", *Scientific Papers on Elasticity and Plasticity*, No. 9, pp. 3–43 (1973) (in Russian).

Chudnovsky, A., "On the Fracture of Solids", in L. Kachanov, ed. *Studies on Elasticity and Plasticity*, Leningrad University Press, Leningrad, USSR, Vol. 9, pp. 2–41 (1973).

Chudnovsky, A., *A Statistical Theory of Brittle Failure*, Transactions of IV Interunion Conference on Theoretical and Applied Mechanics, Kiev (1976) (Abstract, in Russian).

Chudnovsky, A., "Crack Layer Theory", NASA Contractor Report 174634 (1984).

Chudnovsky, A., "Experimental and Theoretical Studies of Slow Crack Growth in Engineering Polymers", *Key Engineering and Mathematics*, Vol. 345, pp. 493–496 (2007).

Chudnovsky, A., "Slow Crack Growth, Its Modeling and Crack-Layer Approach: A Review", *International Journal of Engineering Science*, Vol. 83, pp. 2–4 (2014).

Chudnovsky, A. and Bessendorf, M., "Crack Layer Morphology and Toughness Characterization in Steels", NASA Contractor Report 168154, May 1983.

Chudnovsky, A., Dolgopolsky, A. and Kachanov, M., "Elastic Interaction of a Crack with a Microcrack Array—I. Formulation of the Problem and General form of the Solution", *International Journal of Solids and Structures*, Vol. 23, No. 1, pp. 1–10 (1987a).

Chudnovsky, A., Dolgopolsky, A. and Kachanov, M., "Elastic Interaction of a Crack with a Microcrack Array—II. Elastic Solution for Two Crack Configurations (Piecewise Constant and Linear Approximations)", *International Journal of Solids and Structures*, Vol. 23, pp. 11–21 (1987b).

Chudnovsky, A., Dunaevsky, V. and Khandogin, V., "On the Quasistatic Growth of Cracks", *Archives of Mechanics*, Vol. 30, No. 2, pp. 165–174, Warszawa (1978).

Chudnovsky, A. and Gorelik, M., "Statistical Fracture Mechanics—Basic Concepts and Numerical Realization", in *Book: Probabilities and Materials*, Springer, Netherlands, pp. 321–338 (January 1994)

Chudnovsky, A. and Kachanov, M., "Interaction of a Crack with a Field of Microcracks", *Letters in Applied and Engineering Sciences*, Vol. 21, No. 8, pp. 1009–1018 (1983).

Chudnovsky, A. and Kunin, B., "A Probabilistic Model of Brittle Crack Formation", *Journal of Applied Physics*, Vol. 62, No. 10, p. 4124 (November 1987).

Chudnovsky, A. and Kunin, B., "Statistical Fracture Mechanics", in M. Mareschal and B.L. Holian, eds. *Microscopic Simulation of Complex Hydrodynamic Phenomena*, Plenum Press, New York, Series B: Physics, Vol. 292, pp. 345–360 (1992).

Chudnovsky, A., Kunin, B. and Gorelik, M., "Modeling of Brittle Fracture Based on the Concept of Crack Trajectory Ensemble", *Engineering Fracture Mechanics*, Vol. 58, No. 5–6, pp. 437–457 (November 1997).

Chudnovsky, A. and Moet, A., "A Theory for Translational Crack Layer Propagation", Proceedings of the 6th International Conference on Fracture (ICF6), New Delhi, India, pp. 871–879 (December 4–10, 1984).

Chudnovsky, A., Moet, A., Bankert, R.J. and Takemori, M., "Effect of Damage Dissemination on Crack Propagation in Polypropylene", *Journal of Applied Physics*, Vol. 54, No. 10, pp. 5562–5567 (October 1983).

Chudnovsky, A. and Perdikaris, P., "New Approach to the Fracture Toughness of Concrete—Probabilistic Model", Proceedings of the Fourth International Conference on Applications of Statistics and Probability in Solid and Structural Engineering, University of Florence, Italy, Vol. 1, p. 407 (1983).

Chudnovsky, A., Saada, A. and Lesser, A.J., "Micro Mechanisms of Deformation in Fracture of Over Consolidated Clays", *Canada Geotechnology Journal*, Vol. 25, pp. 213–221 (1988).

Chudnovsky, A. and Shulkin, Y., "Application of Crack Layer Theory to Modeling of Slow Crack Growth in Polyethylene", *International Journal of Fracture*, Vol. 97, pp. 83–102 (1999).

Chudnovsky, A., Zhou, Z., Haiying, Z. and Sehanobish, K., "Lifetime Assessment of Engineering Thermoplastics", *International Journal of Engineering Sciences*, Vol. 59, pp. 108–139 (2012).

Chudnovsky, A.I., *Research on Elasticity and Plasticity* (L.M. Kachanov, ed.), Leningrad University Press, St. Petersburg, Leningrad. In chief (1973).

Curtin, W.A., "Size Scaling of Strength in Heterogeneous Materials", *Physical Review Letters*, Vol. 80, pp. 1445–1448 (1998).

Davidenkov, N., Shevandin, E. and Wittmann, F., "The Influence of Size on the Brittle Strength of Steel", *Journal of Applied Mechanics*, Vol. 14, pp. 63–67 (1947).

de Groot, S.R. and Mazur, P., *Non-Equilibrium Thermodynamics*, North-Holland Publishing Company, Amsterdam, London (1969).

Dugdale, D.S., "Yielding of Steel Sheets Containing Slits", *Journal of the Mechanics and Physics of Solids*, Vol. 8, pp. 100–108 (1960).

Elber, W., "Fatigue Crack Closure Under Cyclic Tension", *Engineering Fracture Mechanics*, Vol. 2, Issue 1, pp. 37–44 (1970).

Elber, W., *The Significance of Crack Closure*, Damage Tolerance in Aircraft Structures, ASTM STP 486, ASTM, Philadelphia, PA (1971)]. Phenomenological modeling of the crack growth retardation was proposed by Wheeler (1972).

Erdogan, E., "Fracture Mechanics", *International Journal of Solids and Structures*, Vol. 37, pp. 171–183 (2000).

Erdogan, F. and Ratwani, M., "Fatigue and Fracture of Cylindrical Shells Containing a Circumferential Crack", *International Journal of Fracture Mechanics*, Vol. 6, pp. 379–392 (1970).

Eshelby, J.D., "The Force on an Elastic Singularity", *Philosophical Transactions of the Royal Society A: Mathematical, Physical and Engineering Sciences*, Vol. 244, No. 877, pp. 87–112 (1951).

Forman, R.G., Kearney, V.E. and Engle, R.M., "Numerical Analysis of Crack Propagation in Cyclic Loaded Structures", *Journal of Basic Engineering, Transaction of the ASME*, Vol 89, pp. 459–464 (1967).

Gefland, I.M. and Jaglom, A.M., "Evaluation of Functional Integrals", *Russian Mathematical Surveys*, Vol. 11, No. 48 (1956).

Glansdorff, P. and Prigogine, I., *Thermodynamic Theory of Structure, Stability and Fluctuations*, Wiley-Interscience, New York (1971).

Glenn, L.A., Gommerstadt, B.Y. and Chudnovsky, A., "A Fracture Mechanics Model of Fragmentation", *Journal of Applied Physics*, Vol. 60, No. 3, pp. 1224–1226 (1986).

Goldenblat, I.I., Bazanov, B.L. and Kopnov, V.A., *Long Term Strength in the Machine Building, Machine Building* (1977).

Golubovic, L. and Feng, S., "Rate of Microcrack Nucleation", *Physical Review A*, Vol. 43, pp. 5223–5227 (1991).

Grellman, W. and Seidler, S. (eds.), *Deformation and Fracture Behavior of Polymers*, Springer-Verlag, Heidelberg, Germany (2001).

Griffith, A.A., "The Phenomena of Rupture and Flow in Solids", *Philosophical Transactions of the Royal Society of London A*, Vol. 221, pp. 163–197 (1921).

Griffith, A.A., "The Theory of Rupture", in C.B. Biezeno and J.M. Burgers, eds. *Proceedings, First International Congress for Applied Mechanics*, J. Waltman, jr., Delft, pp. 55–63 (1925).

Haase, R., *Thermodynamic Der Irreversiblen Prozesse*, Dr. Dietrich Steinkopf Verlag, Darmstadt (1963) (in German).

Hashida, T. and Takahashi, H., "Significance of AE Crack Monitoring in Fracture Toughness Evaluation and Non-linear Rock Fracture Mechanics", *International Journal of Rock Mechanics and Mining Sciences & Geomechanics Abstracts*, Vol. 30, Issue 1, pp. 47–60 (1993).

Hill, R., *The Mathematical Theory of Plasticity*, Oxford University Press, London, chapter 12, p. 323 (1950).

Hoagland, R.G.Hahn, G.T. and Rosenfield, A.R., "Influence of Microstructure on Fracture Propagation in Rock", *Rock Mechanics Felsmechanik Mecanique des Roches*, Vol. 5, pp. 77–106 (1975).

Huang, Y.-L. and Brown, N., "The Effect of Molecular Weight on Slow Crack Growth in Linear Polyethylene Homopolymers", *Journal of Materials Science*, Vol. 23, pp. 3648–3655 (1988).

Inglis, C.E., "Stresses in Plates Due to the Presence of Cracks and Sharp Corners", *Transactions of the Institute of Naval Architects*, Vol. 55, pp. 219–241 (1913).

Irwin, G.R., "Analysis of Stresses and Strains Near the End of a Crack Traversing a Plate", *Transaction ASME, Journal of Applied Mechanics*, Vol. 24, pp. 361–364 (1957).

Irwin, G.R., *Fracture Handbuch der Physik*, Springer-Verlag, Heidelberg, Vol. VI, pp. 551–590 (1958).

Ising, E., "Beitrag zur Theorie des Ferromagnetismus", *Zeitschrift für Physik* (in German), Vol. 31, pp. 253–258 (1925).

Issa, M.A., Hammad, A.M. and Chudnovsky, A., "Correlation between Crack Tortuosity and Fracture Toughness in Cementitious Material", *International Journal of Fracture*, Vol. 60, pp. 97–105, Kluwer Academic Publishers, Printed in the Netherlands (1993).

Jasarevic, H., Chudnovsky, A., Dudley, J.W. and Wong, G.K., "Observation and Modelling of Brittle Fracture Initiation in a Micro-Heterogeneous Material", *International Journal of Fracture*, Letters in Fracture and Micromechanics, pp. 73–80 (June 23, 2009).

Kachanov, L.M., "On Time to Failure Under Creep Conditions", *News AS (Academy of Science) USSR, Technology Science Department*, No. 8 (1958) (in Russian).

Kachanov, L.M. and Montagut, E., "Interaction of Crack with Certain Microcrack Arrays", *Engineering Fracture Mechanics*, Vol. 25, No. 5–6, pp. 625–636 (1986).

Kanuan, S. and Chudnovsky, A., "A Model of Quasibrittle Fracture of Solids", *International Journal of Damage Mechanics*, Vol. 8, pp. 19–40 (1999).

Katz, Victor J., "The History of Stokes's Theorem", *Mathematics Magazine*, Vol. 52, No. 3, pp. 146–156 (1979).

Khandogin, V. and Chudnovsky, A., "Thermodynamics of Quasistatic Growth", in *The Book Dynamics and Strength of Aircraft Constructions*, Novosibirsk, Vol. 4, pp. 148–175 (1978) (in Russian).

Kirsch, E.G., "Die Theorie der Elastizität und die Bedürfnisse der Festigkeitslehre", *Zeitschrift des Vereines deutscher Ingenieure*, Vol. 42, pp. 797–807 (1898).

Kiyalbaev, D.A. and Chudnovsky, A.I., "On the Nature of Failure", Mechanics of Lattice Systems and Continuous Media, Collection: Tr. Leningr. Inzh. -Stroit, In-ta, No. 60 (1969) [Trans. Of the Leningrad Civil Engineering Inst.].

Knott, J.F., *Fundamentals of Fracture Mechanics*, Butterworth, Seven Oaks (1973).

Knowles, J.K. and Sternberg, E., "On a Class of Conservation Laws in Linearized and Finite Elasto-Statics", *Archive for Rational Mechanics and Analysis*, Vol. 44, No. 3 (1972).

Krajcinovic, D. and Fonseka, G., "The Continuous Damage Theory of Brittle Materials", *ASME Journal of Applied Mechanics*, Vol. 48, pp. 809–815 (1981).

Kramer, E.J., "Craze Fibril Formation and Breakdown", *Polymer Engineering and Science*, Vol. 24, No. 10, pp. 761–769 (July 1984).

Kunin, B. and Gorelik, M., "On Representation of Fracture Profiles by Fractional Integrals of a Wiener Process", *Journal of Applied Physics*, Vol. 70, No. 12, pp. 7651–7653 (1991).

Lavoie, J.L., Osler, T.J. and Tremblay, R., "Fractional Derivatives and Special Functions", *SIAM Review*, Vol. 18, No. 2, pp. 240–268 (1976).

Leevers, P.S., Yayla, P. and Wheel, M.A., "Charpy and Dynamic Fracture Testing for Rapid Crack Propagation in Polyethylene Pipe", *Plastics, Rubber and Composites Processing and Applications*, Vol. 17, pp. 247–253 (1992).

Lemaitre, J. and Chaboche, J.L., *Mechanics of Solid Materials*, Cambridge University Press, Cambridge (2002).

Lesser, A., "Theoretical and Experimental Studies of Cooperative Fracture in Over Consolidated Clays", Dissertation for Doctor of Philosophy, Advisors: Dr. Adel Saada and Dr. Alexander Chudnovsky, Case Western Reserve University, January 1989.

Liu, J., Zhou, Z., Niu X. and Chudnovsky, A., "True-Stress–Strain-Temperature Diagrams of Polyolefins and Their Application in Acceleration Tests for Lifetime Predication", Proceedings of the 58th Annual Technical Conference & Exhibition, ANTEC 2000, Society of Plastics Engineers, Orlando, FL, pp. 3189–3193 (2000).

Lo, K.K., "Analysis of Branched Cracks", *Journal of Applied Mechanics*, Vol. 45, pp. 797–802 (1978).

Lund, J.R. and Byrne, J.P., "Leonardo Da Vinci's Tensile Strength Tests: Implications for the Discovery of Engineering Mechanics", *Civil Engineering and Environmental Systems*, pp. 1–8 (2000).

Mandelbrot, B.B., "How Long is the Coast of Britain? Statistical Self-Similarity and Fractional Dimension", *Science*, Vol. 156, No. 3775, pp. 636–638 (1967).

Mandelbrot, B.B., Passoja, D.E. and Paullay, A., "Fractal Character of Fracture Surfaces of Metals", *Nature*, Vol. 308, pp. 721–722 (1984).

Mostovoy, S., Crosley, P.B. and Ripling, E.J., "Use of Crack-Line-Loaded Specimens for Measuring Plane-Strain Fracture Toughness", *Journal of Materials*, Vol. 2, pp. 661–681 (1967).

Mull, M.A., Chudnovsky, A. and Moet, A., "A Probabilistic Approach to the Fracture Toughness of Composites", *Philosophical Magazine A*, Vol. 56, No. 3, pp. 419–443 (1987).

Onsager, L., "Reciprocal Relations in Irreversible Processes", ii., *Physical Review*, Vol. 38, pp. 2265–2279 (1931).

Paris, P.C. and Erdogan, F., "A Critical Analysis of Crack Propagation Laws", *Transactions of the American Society of Mechanical Engineers*, Vol. 85, pp. 528–534 (December 1963).

Parker, A.P., *The Mechanics of Fracture and Fatigue: An Introduction*, Spon publisher, London, New York (1981).

Parsons, M., Stepanov, E.V., Hiltner, A. and Baer, E., "Correlation of Stepwise Fatigue and Creep Slow Crack Growth in High Density Polyethylene", *Journal of Materials Science*, Vol. 34, p. 3315 (1999).

Parsons, M., Stepanov, E.V., Hiltner, A. and Baer, E., "The Damage Zone Ahead of the Arrested Crack in Polyethylene Resins", *Journal of Materials Science*, Vol. 36, p. 5747 (2001).

Patel, R.M., Sehanobish, K., Jain, P., Chum, S.P. and Knight, G.W., "Theoretical Prediction of Tie-Chain Concentration and its Characterization Using Post-Yield Response", *Journal of Applied Polymer Science*, Vol. 60, No. 5, pp. 749–758 (May 2, 1996).

Perzyna, P., "Fundamental Problems in Viscoplasticity", *Advances in Applied Mechanics*, Vol. 9, No. 2, pp. 244–368 (1966).

Prigogine, I., *Introduction to Thermodynamics of Irreversible Processes*, 3rd edition, Wiley Interscience, New York (1955/1961/1967).

Prigogine, I. and Defay, R., *Chemical Thermodynamics*, Longmans, Green & Co, London (1950/1954).

Ravi-Chandar, K. and Knauss, W.G., "An Experimental Investigation into Dynamic Fracture: III. On Steady-State Crack Propagation and Crack Branching", *International Journal of Fracture*, Vol. 26, pp. 141–154 (October 1984).

Rice, J.R., "A Path Independent Integral and the Approximate Analysis of Strain Concentration by Notches and Cracks", *Journal of Applied Mechanics, Transaction*. ASME, Series E, Vol. 35, pp. 379–386 (June 1968).

Rice, J.R. and Thompson, R., "On the Ductile Versus Brittle Behavior of Crystals", *Philosophical Magazine*, Vol. 29, pp. 73–97 (1974).

Rosin, P. and Rammler, E., "The Laws Governing the Fineness of Powdered Coal", *Journal of Institute of Fuel*, Vol. 7, pp. 29–36 (1933).

Schicker, J. and Pfuff, M., "Statistical Modeling of Fracture in Quasi-Brittle Materials", *Advanced Engineering Materials*, Vol. 8, pp. 406–410 (2006).

Slepjan, L.I., *Mechanics of Cracks*, 2nd edition, Ship Building, Leningrad (1990).

Suzuki, M., Abe, H., Takanashi, H., Tamakawa, K. and Kikuchi, M., "Acoustic Emission Characteristics and Fracture Toughness of Sandstone", *Technical Reports of The Tohoku University*, Vol. 45, pp. 239–272 (1980).

Swalin, R., *Thermodynamics of Solids*, 2nd edition, Wiley-Interscience Publication, John Wiley & Sons, New York, Chichester, Brisbane, Toronto, pp. 76–81 (1972).

Tada, H., Paris, P.C. and Irwin, G.R., *The Stress Analysis of Cracks Handbook*, 2nd edition. Paris Productions Inc., St. Louis (1985).

Tetsuya, K., Hisashi, A., Akira, M. and Yukinori, M., "Effect of Specimen Size and Rock Properties on the Uniaxial Compressive Strength of Ryukyu Limestone", *Journal of Japan Society of Engineering Geology*, Vol. 46, pp. 2–8 (2005).

Timoshenko, S., *Strength of Materials—Part II Advanced Theory and Problems*, Robert E. Krieger Publishing, New York, Reprint 1976.

Timoshenko, S., *History of Strength of Materials*, McGraw-Hill, New York (1953).

Timoshenko, S. and Woinowsky-Krieger, S., *Theory of Plates and Shells*, McGraw—Hill, New York (1959).

Todhunter, I. and Pearson, K., *A History of the Theory of Elasticity*, University Press, Dover Publications, Cambridge (1886).

Tresca, H. Sur, "l'écoulement des corps solides soumis â de fortes pressions", *Comptes rendus de l'Académie des Sciences, Paris*, Vol. 59, p. 754 (1864).

Von Mises, R., *Mechanik der festen Körper im plastisch deformablen Zustand*, Nachr Ges Wiss, Göttingen, p. 582 (1913).

Walker, K., *The Effect of Stress Ratio During Crack Propagation and Fatigue for 2024-T3 and 7075-T6 Aluminum*, Effects of Environment and Complex Load History on Fatigue Life, ASTM, Philadelphia, pp. 1–14 (1970).

Weibull, W., "A Statistical Theory of the Strength of Materials", *Royal Swedish Institute for Engineering Research*, No. 151 (1939).

Weibull, W., "A Statistical Distribution Function of Wide Applicability", *Journal of Applied Mechanics*, Vol. 18, pp. 293–297 (1951).

Westergaard, H.M., "Bearing Pressures and Cracks", *Journal of Applied Mechanics*, Vol. 6, pp. 49–53 (1939).

Wheeler, O.E., "Spectrum Loading and Crack Growth", in *Proceedings of the International Conference on Fatigue in Metals*, Institute of Mechanical Engineers, London, pp. 531–554 (1972).

William Parsons, B., *Engineers and Engineering in the Renaissance*, MIT Press, Cambridge, MA, p. 661 (1939).

Willis, J.R., "A Comparison of the Fracture Criteria of Griffith and Barenblatt", *Journal of The Mechanics and Physics of Solids*, Vol. 15, pp. 151–162 (1967).

Wu, S. and Chudnovsky, A., "Elastic Interaction of a Crack with a Random Array of Microcracks", *International Journal of Fracture*, Vol. 49, pp. 123–140 (1991).

Zhang, H. and Chudnovsky, A., "Slow Crack Growth in High Density Polyethylene Part II: Simulations Using Crack Layer Model", *SPE-ANTEC Tech Papers*, Vol. 59, p. 1295 (2013).

Zhang, H., Chudnovsky, A., Wong, G. and Dudley, J.W., "Statistical Aspects of Micro Heterogeneous Rock Fracture: Observations and Modelling", *Rock Mechanics and Rock Engineering*, Vol. 46, No. 3, pp. 4999–4514 (2013).

Zhang, H., Zhou, Z. and Chudnovsky, A., "Applying the Crack Layer Concept to Modelling of Slow Crack Growth in Polyethylene", *International Journal of Engineering Science*, Vol 83, p. 42 (2014).

Zhou, Z., Caratus, A., Michie, W. and Chudnovsky, A., "Characterization of Slow Crack Growth in HDPE Under Creep Condition", Proceedings of the 66th Annual Technical Conference & Exhibition, ANTEC 2008, Society of Plastics Engineers, Milwaukee, WI, pp. 746–751 (2008).

Zhou, Z., Chen, D., Chudnovsky, A., Shulkin, Y., Jivraj, N. and Sehanobish, K., "Ductile Failure and Delayed Necking in Polyethylene", Proceedings of the 58th Annual Technical Conference & Exhibition, ANTEC 2000, Society of Plastics Engineers, Orlando, FL, pp. 3148–3152 (2000).

Zhou, Z., Chudnovsky, A., Bosnyak, C.P. and Sehanobish, K., "Cold-Drawing (Necking) Behavior of Polycarbonate as a Double Glass Transition", *Polymer Lifetime Engineering Science*, Vol. 35, pp. 304–309 (1995).

Zhou, Z., Chudnovsky, A. and Sehanobish, K., "Evaluation of Time to Ductile Failure in Creep of PEs from Short-Term Testing", Proceedings of the 63rd Annual Technical Conference & Exhibition, ANTEC 2005, Society of Plastics Engineers, Boston, MA, pp. 784–791 (2005).

Zhou, Z., Chudnovsky, A., Sehanobish, K. and Bosnyak, C.P., "The Time Dependency of the Necking Process in Polyethylene", Proceedings of the 57th Annual Technical Conference & Exhibition, ANTEC 1999, Society of Plastics Engineers, New York, NY, pp. 3399–3402 (1999).

Zhou, Z., Chudnovsky, A., Michie, W. and Demirors, M., "Prediction of PE Long-Term Creep Property and Lifetime in Ductile Failure Based on Short-Term Tests, Proceedings of the 68th Annual Technical Conference & Exhibition, ANTEC 2010, Society of Plastics Engineers, Orlando, FL, pp. 664–668 (2010).

Zhou, Z., Zhang, H. and Chudnovsky, A., "Temperature Effects on Slow Crack Growth in Pipe Grade PE", Proceedings of the 68th Annual Technical Conference & Exhibition, ANTEC 2010, Society of Plastics Engineers, Orlando, FL, pp. 679–684 (2010).

Zhou, Z., Zhang, H., Chudnovsky, A. and Sehanobish, K., "Slow Crack Growth in High Density Polyethylene Part I: Experimental Observations", *SPE-ANTEC Tech Papers*, Vol. 59, p. 1290 (2013).

Appendix I

Summary of Onsager Reciprocal Relations: Lars Onsager used fluctuation theory to find reciprocal relations among the transport coefficients. The Onsager reciprocal relation connects thermodynamics, transport theory, and statistical mechanics. It specifically plays an important role in nonequilibrium thermodynamics.

Onsager used a notation J1 for the electric current and J2 for heat flow, recognizing the current is driven by electromotive force, which he called X1. In corresponding units, the "force" which drives the flow of heat will be:

$$X_2 = -\frac{1}{T} \, gradT,$$

where T denotes the absolute temperature. If the heat flow and the current were completely independent, he claimed relations of the type:

$$X_1 = R_1 J_1$$
$$X_2 = R_2 J_2$$

where R_1 is the electrical resistance and R_2 is the heat resistance. He also proposed relations when the two processes interfere. For electrical conduction, the relation on the top holds (Ohm's law). Similar relationships between forces and fluxes hold true for empirical relations like Fick's law for diffusion and Darcy's law for filtration. For Fick's law, "force" is proportional to the concentration gradient, and for Darcy's law "force" is proportional to the pressure gradient. This can be also extended to filtration where both forces are operative.

Appendix II

Historical Remarks on Energy Release Rates: J integral was first introduced by Eshelby in 1951 [J.D. Eshelby, "The force on an elastic singularity", Philos. Trans. Roy. Soc. London A, 244, pp 87–112 (1951)] to express the force acting on a singularity within an elastic solid. Later, J integral was rederived independently by Sanders in 1960 [J. L. Sanders, "*On the Griffith-Irvin Fracture Theory*", J. appl. Mech., N2, 1960.], Cherepanov in 1967 [G. P. Cherepanov "*On Mathematical Theory of Cracks at Equilibrium*" MTT, N6, 1967 (in Russian), G.P. Cherepanov, "*Invariant Γ integrals*", Eng. Fract. Mech., 14 (1981), pp. 39–58], and Rice in 1968 [J. R. Rice "*A Path Independent Integral and the Approximate Analysis of Strain Concentration by Notches and Cracks*", J. of Appl. Mech. 35, Trans. ASME, Series E, June 1968]. The most clear and popular interpretation of J as the energy release rate with respect to crack length has been done by Rice. Gunter [W. Gunter, Abh. Braunschw. Wiss. Ges., 14, 54, 1962] and later Knowles and Sternberg [J. K. Knowles, E. Sternberg "*On a Class of Conservation Laws in Linearized and Finite Elastostatics*", Arch. Rational Mech. And Analysis, 44, N3, 1972] applied Noether's theorem to elastostatics and obtained three path-independent integrals for a linear, homogeneous, and isotropic medium integrals J, L, and M associated with translational, rotational, and expansional invariances (see *Appendix III* for remarks on Noether). These integrals express the general conservation laws of elastostatics. Shortly after Knowles and Sternberg's publication, the physical interpretation of L and M integrals were discussed by Budiansky and Rice [B. Budiansky, J. R. Rice "*Conservation Laws and Energy Release Rates*", J. Appl. Mech. Vol. 40, March 1973]. The same J, L, and M integrals (and additional one N) appeared in the Crack layer theory as active parts of thermodynamic forces reciprocal to Crack layer extension, rotation, expansion, and distortion [A. Chudnovsky, V. Dunaevsky, V. Khandogin "*On the Quasistatic Growth of Cracks*", Arch. Mech. 30, 2, pp. 165–174, Warszawa, 1978; V. Khandogin, A. Chudnovsky "*Thermodynamics of Quasistatic Growth*", in the book Dynamics and Strength of Aircraft Constructions, 4, pp. 148–175., Novosibirsk, 1978 (in

Russian)]. In a recent publication of Aoki, Kishimoto, and Sakata [S. Aoki, K. Kishimoto, M. Sakata *"Energy-Release Rate in Elasto-Plastic Fracture Problems"*, J. Appl. Mech. Vol. 48, December 1981], a generalization of *J*, *L*, and *M* integrals is proposed for cases in which plastic deformation, body forces, and thermal strains may exist. The formalism used in [S. Aoki, K. Kishimoto, M. Sakata *"Energy-Release Rate in Elasto-Plastic Fracture Problems"*, J. Appl. Mech. Vol. 48, December 1981] is very similar to that in [A. Chudnovsky, V. Dunaevsky, V. Khandogin *"On the Quasistatic Growth of Cracks"*, Arch. Mech. 30, 2, pp. 165–174, Warszawa, 1978; V. Khandogin, A. Chudnovsky *"Thermodynamics of Quasistatic Growth"*, in the book Dynamics and Strength of Aircraft Constructions, 4, pp. 148–175., Novosibirsk, 1978 (in Russian)]. Path-independent integrals for inelastic materials are in details discussed by Stonesifer and Atluri [R. Stonesifer, S. Atluri *"Creep Crack Growth: A New Path-Independent Integral (I) and Computational Studies"* NASA Report NAG3–38, 1981].

Fractal Dimension: In fractal geometry, a fractal dimension is a ratio providing a statistical index of complexity comparing how details in a fractal pattern change with the scale at which it is measured. It has also been characterized as a measure of the space-filling capacity of a pattern that tells how a fractal scales differently from the space it is embedded in; a fractal dimension does not have to be an integer.

The essential idea of "fractal" dimensions has a long history in mathematics, but the term itself was brought to the forefront by Benoit Mandelbrot on the basis of his 1967 paper on self-similarity in which he discussed fractional dimensions [Mandelbrot (1967)]. In that paper, Mandelbrot cited previous work by Lewis Fry Richardson describing the counter-intuitive notion that a coastline's measured length changes with the length of the measuring stick used. In terms of that notion, the fractal dimension of a coastline quantifies how the number of scaled measuring sticks required to measure the coastline changes with the scale applied to the stick. There are several formal mathematical definitions of fractal dimensions that build on this basic concept of change in detail with change in scale. Ultimately, the term fractal dimension became the phrase with which Mandelbrot himself became most comfortable with respect to encapsulating the meaning of the word fractal, a term he created. After several iterations over the years, Mandelbrot settled on this use of the language.

The concept of a fractal dimension rests in unconventional views of scaling and dimension. Traditional notions of geometry dictate that shapes scale predictably according to intuitive and familiar ideas about the space they are contained within, such that, for instance, measuring a line using first one measuring stick then another one-third its size, will give for the second stick a total length three times as many sticks long as with the first. This is held in two dimensions, as well. If one measures the area of a square and then measures again with a box of side length one-third the size of the original, then one will find nine times as many squares as with the first measure. Such familiar scaling relationships can be defined mathematically by the general scaling rule in Eq. (1), where the variable N stands for the number of measurement units (sticks, squares, etc.) for the scaling factor, and D for the fractal dimension:

$$N = \varepsilon^{-D} \tag{1}$$

This scaling rule typifies conventional rules about geometry. $D = 1$ for lines because $N = 3$ when $\varepsilon = 1/3$. Many real-world phenomena exhibit limited or statistical fractal properties and fractal dimensions. Practically, measurements of fractal dimension are affected by various methodological issues and are sensitive to numerical or experimental noise and limitations in the amount of data. Nonetheless, the field is rapidly growing as estimated fractal dimensions for statistically self-similar phenomena may have many practical applications in various fields including astronomy, acoustics, geology and earth sciences, diagnostic imaging, ecology, electrochemical processes, image analysis, biology and medicine, neuroscience, network analysis, physiology, and physics.

Appendix III

Remarks on Emmy Noether's theorem: Noether's first theorem states that every differentiable symmetry of the action of a physical system with conservative forces has a corresponding conservation law. The theorem was proven by mathematician Emmy Noether in 1915 and published in 1918 [Noether E (1918). "Invariante Variationsprobleme". Nachr. D. König. Gesellsch. D. Wiss. Zu Göttingen, Math-phys. Klasse, pp 235–257, 1918], after a special case was proven by E. Cosserat and F. Cosserat in 1909. The action of a physical system is the integral over time of a Lagrangian function, from which the system's behavior can be determined by the principle of least action. This theorem only applies to continuous and smooth symmetries over physical space. Noether's theorem is used in theoretical physics and the calculus of variations.

Noether's second theorem relates symmetries of an action functional with a system of differential equations. The action S of a physical system is an integral of a so-called Lagrangian function L, from which the system's behavior can be determined by the principle of least action.

Noether's second theorem is sometimes used in gauge theory. Gauge theories are the basic elements of all modern field theories of physics, such as the prevailing Standard Model.

Amalie Emmy Noether (who lived from 23 March 1882 to 14 April 1935) was a German mathematician who made many important contributions to abstract algebra. She discovered Noether's theorem, which is fundamental in mathematical physics. She invariably used the name "Emmy Noether" in her life and publications. She was described by Albert Einstein, Norbert Wiener, and several others as the most important woman in the history of mathematics. As one of the leading mathematicians of her time, she developed some theories of rings, fields, and algebras. In physics, Noether's theorem explains the connection between symmetry and conservation laws.

Noether was born to a Jewish family; her father was the mathematician, Max Noether. She originally planned to teach French and English after passing the required examinations, but instead studied mathematics at the University of Erlangen, where her father lectured. After completing her doctorate in 1907 under the supervision of Paul Gordan, she worked at the Mathematical Institute of Erlangen without pay for seven years. At the time, women were largely excluded from academic positions. In 1915, she was invited by David Hilbert and Felix Klein to join the Mathematics Department at the University of Göttingen, a world-renowned center of mathematical research. The philosophy faculty objected, and she spent four years lecturing under Hilbert's name with a special permission.

Appendix IV

Configurational versus Thermodynamic Forces: The notion of forces is central to all of continuum mechanics. Classically, the response of a body to deformation is described by well-understood Newtonian forces. Eshelby was the first one to recognize that additional *configurational forces* may be needed to describe phenomena associated with the material itself, which was followed by Gibb's discussion of multiphase equilibria. Based on Cahn's interpretation of Gibb's discussion,

> Solid surfaces can have their physical area changed in two ways, either by creating or destroying surface without changing surface structure and properties per unit area, or by an elastic strain . . . along the surface keeping the number of surface lattice sites constant.

The creation of surface involves configurational forces, while stretching the surface involves standard forces.

Thermodynamic forces appear in nonequilibrium thermodynamics. A profound difference separates equilibrium from nonequilibrium thermodynamics. Equilibrium thermodynamics ignores the time courses of physical processes. In contrast, nonequilibrium thermodynamics attempts to describe their time courses in continuous. A thermodynamic driving force occurs when a difference in potential exists. Depending on the system, these potentials could be chemical potential (for the transport of the substance between regions), electric potential (associated with current), and others.

Index

Printed in the United States
by Baker & Taylor Publisher Services